Gilbert Hector
Ulrich Hirsch

Introduction to the Geometry of Foliations, Part A

Aspects of Mathematics

Aspekte der Mathematik

Editor: Klas Diederich

The texts published in this series are intended for graduate students and all mathematicians who wish to broaden their research horizons or who simply want to get a better idea of what is going on in a given field. They are introductions to areas close to modern research at a high level and prepare the reader for a better understanding of research papers. Many of the books can also be used to supplement graduate course programs.

The series will comprise two sub-series, one with English texts only and the other in German.

Gilbert Hector
Ulrich Hirsch

Introduction to the Geometry of Foliations, Part A

Foliations on Compact Surfaces, Fundamentals for Arbitrary Codimension, and Holonomy

V

Friedr. Vieweg & Sohn Braunschweig/Wiesbaden

CIP-Kurztitelaufnahme der Deutschen Bibliothek

Hector, Gilbert:
Introduction to the geometry of foliations / Gilbert Hector; Ulrich Hirsch. – Braunschweig; Wiesbaden: Vieweg
(Aspects of mathematics)

NE: Hirsch, Ulrich:

Pt. A. → Hector, Gilbert: Foliations on compact surfaces, fundamentals for arbitrary codimension, and holonomy

Hector, Gilbert:
Foliations on compact surfaces, fundamentals for arbitrary codimension, and holonomy / Gilbert Hector; Ulrich Hirsch. – Braunschweig, Wiesbaden: Vieweg, 1981.
(Introduction to the geometry of foliations / Gilbert Hector, Ulrich Hirsch; Pt. A)
(Aspects of mathematics; 1)
ISBN 3-528-08501-0

NE: Hirsch, Ulrich:; Aspects of mathematics / E

Dr. *Gilbert Hector* is Professor of Mathematics at the Université des Sciences et Techniques de Lille I, France.

Dr. *Ulrich Hirsch* is Privatdozent at the Faculty of Mathematics at the University of Bielefeld, Germany.

1981

Produced by Lengericher Handelsdruckerei, Lengerich
Printed in Germany

ISBN 3-528-08501-0

PREFACE

Foliation theory grew out of the theory of dynamical systems on manifolds and Ch. Ehresmann's connection theory on fibre bundles. Pioneer work was done between 1880 and 1940 by H. Poincaré, I. Bendixson, H. Kneser, H. Whitney, and W. Kaplan - to name a few - who all studied "regular curve families" on surfaces, and later by Ch. Ehresmann, G. Reeb, A. Haefliger and others between 1940 and 1960. Since then the subject has developed from a collection of a few papers to a wide field of research. Nowadays, one usually distinguishes between two main branches of foliation theory, the so-called quantitative theory (including homotopy theory and characteristic classes) on the one hand, and the qualitative or geometric theory on the other.

The present volume is the first part of a monograph on *geometric* aspects of foliations. Our intention here is to present some fundamental concepts and results as well as a great number of ideas and examples of various types. The selection of material from only one branch of the theory is conditioned not only by the authors' personal interest but also by the wish to give a systematic and detailed treatment, including complete proofs of all main results. We hope that this goal has been achieved.

Our exposition is devided into three chapters. In chapter I we study foliations on compact surfaces. This is because, on surfaces, a great number of notions and phenomena which are also relevant to foliations on arbitrary manifolds can be described in a particularly accessible way. For instance, different leaf types and minimal sets can be easily visualized, holonomy is particularly simple, the structurally stable foliations are well known, and even a topological classification of all foliations is possible.

Although foliations on compact surfaces meanwhile belong to the classical part of foliation theory (here essential work was already done by Poincaré and others in the last third of the last century) some of our contributions may be considered as original. Thus our proofs of Kneser's existence theorem for topological foliations and of his compact leaf theorem on the Klein bottle, though completely elementary, are considerably easier than Kneser's original proofs in [Kn]. (Letters in parenthesis refer to the bibliography at the end of the book).

In chapter II, we develop progressively the general notion of a foliation, beginning with foliated bundles (roughly, fibre bundles with a transverse foliation). The holonomy representation of foliated bundles is studied in detail; it will serve us in chapter III as a "model" for the holonomy in general. Besides this, we describe foliations which are defined by a Lie group action. Moreover, the relation between foliations and plane bundles over the underlying manifold is clarified.

The third chapter treats holonomy, which is, without doubt, the central concept in the geometric theory. Here our approach might seem somewhat too detailed and formal than is necessary for the purposes at hand. We took this approach, however, because of the following three advantages:

- It clarifies the development of the previously introduced holonomy for foliated bundles into a generalized holonomy for arbitrary foliations.
- Our definition of holonomy via "unwrapping" the foliation in the neighbourhood of a leaf enables us to consider only properly embedded leaves which are more easily visualized.
- The proofs of two important (possibly the most important) results of the early days of foliation theory, namely Reeb's local stability theorem and Haefliger's theorem stating that holonomy characterizes the foliation in the neighbourhood of a proper leaf, become particularly transparent.

Two further volumes are planned. Part B will deal exclusively with codimension one foliations; its list of contents will probably include the following topics: Fundamentals on codimension one foliations, foliations on spheres, exceptional minimal sets, invariant measures, ends, growth, foliations without holonomy. Part C will treat 2-dimensional foliations on 3-manifolds, including construction principles, Novikov's compact leaf theorem, foliations on Seifert manifolds, foliations on 3-manifolds with solvable fundamental group, foliations defined by $\mathbb{R}^2$-actions, analytic foliations, topological types of proper leaves.

As for prerequisites, the book does not require any preliminary knowledge of foliations. In particular, chapter I is completely elementary and can be read without further reference by anybody who has attended, say, a one year course in analysis and topology. In the second and third chapter, however, some familiarity with differential topology and differential geometry, including vector bundles and Lie groups, is desirable. We have tried to make the text as self-contained as possible, but in certain cases where some general material is needed we refer the reader to the literature.

The exercises are meant to provide practice and familiarity with the concepts of the main text. There should not be any unsolved problems among them.

The symbol □ is used to indicate the end of a proof. Items are numbered consecutively, and the reference II; 2.1.1 refers to item 2.1.1 in the second chapter. Items within a chapter are cited simply as, say, 2.1.1. A summary of basic notations used throughout this text can be found at the end of the book after the bibliography.

In concluding this preface the authors express their gratitude to D. Zagier and W.D. Neumann who read most parts of the manuscript with great care. Their suggestions have led to many improvements in the text. Thanks are also due to I. Lieb and S. Morita for discussions and to the Secretariat

Scientifique de l'UER de Mathematiques de Lille for typing the main portion of the manuscript. The second author also thanks Heinrich-Hertz-Stiftung of Nordrhein Westfalen government for financial support during the preparation of this text. Last, but not least, the authors are grateful to both the editor K. Diederich and Vieweg Verlag for offering them the opportunity of beginning a new mathematical series with their contribution.

Finally, we invite the readers to communicate their comments on this volume to us.

G. Hector and U. Hirsch

CHAPTER I - FOLIATIONS ON COMPACT SURFACES.

CHAPTER I

FOLIATIONS ON COMPACT SURFACES.

1. Vector fields on surfaces.

The theory of foliations has one of its roots in the study of the solution curves of ordinary differential equations on $\mathbb{R}^2$ or, more generally, of vector fields on surfaces. This is one reason why we start our investigations of foliations on manifolds with vector fields on surfaces. Another reason for this approach is the fact that many of the phenomena on manifolds of higher dimensions which will be studied in this book already appear on surfaces and can be most easily described there.

1.1. Examples of vector fields on surfaces.

Although foliations will not have "singularities" it will turn out that vector fields with singularities will be useful, for example, to characterize those surfaces which admit foliations. Therefore we begin with a list of some examples of vector fields which are defined in a neighbourhood of $0 \in \mathbb{R}^2$ and which have 0 as an isolated singularity. Here vector fields are considered as C^r maps $X : U \to \mathbb{R}^2$, where $1 \leq r \leq \infty$, and U is an open neighbourhood of 0 in $\mathbb{R}^2$. (The reader not familiar with vector fields, flows, etc. is referred to Sternberg [Ste], for example.

a) $X(x,y) = (x,y)$,

with flow lines starting from 0; see figure 1a).

b) $X(x,y) = (-x,-y)$,

with flow lines running to 0; see figure 1b).

c) $X(x,y) = (-y,x)$

with concentric circles as flow lines. This type of singularity is usually referred to as a <u>center</u> of X, cf. fig. 1c).

d) $X(x,y) = (x,-y)$,

with hyperbolic flow lines, fig. 1d). A singularity of this type is called a saddle (point). For the motivation for this name, cf. fig. 2i).

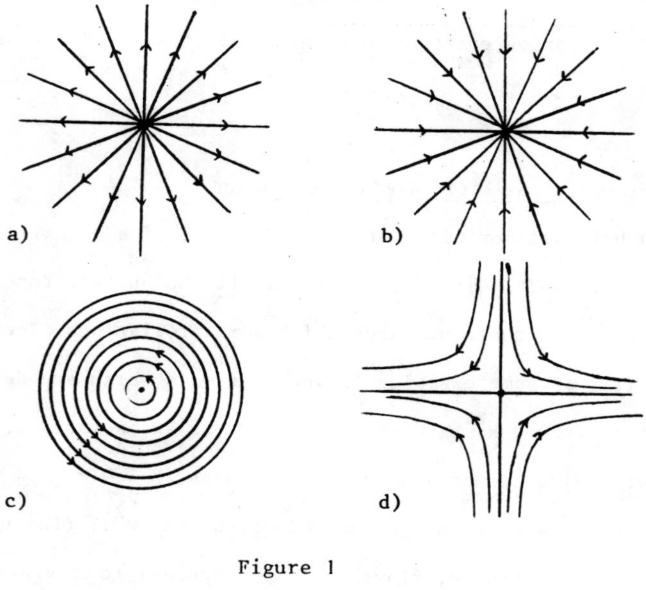

Figure 1

Here are some examples of vector fields globally defined on surfaces. We dispense with a precise definition and only indicate the orbits.

e) A flow on the 2-sphere S^2 with parallel circles as orbits and north and southpole as singularities of center type; see fig. 2e).

f) A flow on S^2 with a single singularity, fig. 2f).

g) Consider $\mathbb{R}^2$ as the universal covering of the 2-torus T^2. For every $\alpha \in \mathbb{R}$ the vector field $X_\alpha = id_{\mathbb{R}^2} + (1,\alpha)$ on $\mathbb{R}^2$ is invariant under covering translations and therefore induces a vector field on T^2. The flow lines of X_α are the lines of slope α. If α is rational then all flow lines are closed curves, fig. 2g), (1) and (2). Otherwise they are all homeomorphic to $\mathbb{R}$.

h) A flow on the Möbius band B tangent to the boundary and without singularities, see fig. 2h). Doubling B gives a flow on the Klein bottle, also without singularities.

i) A flow on the 2-sphere with three holes (this surface is sometimes called a pair of pants), transverse to the boundary and with one singularity which is of saddle type, fig. 2i), (1) and, equivalently, fig. 2i), (2).

Figure 2

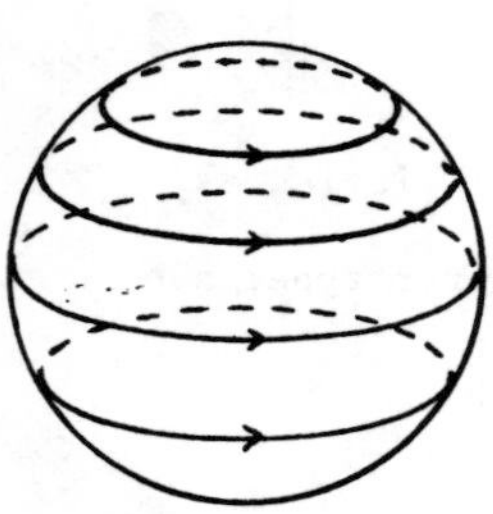

e)

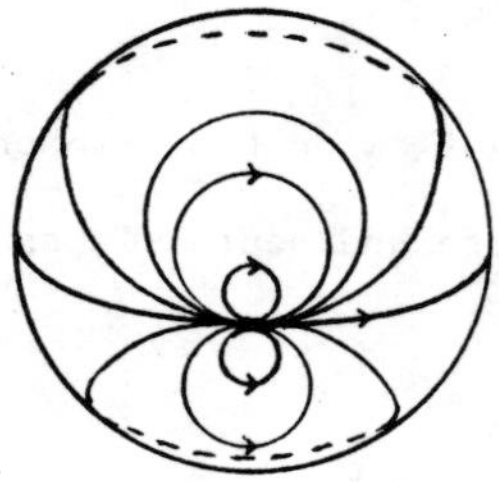

f)

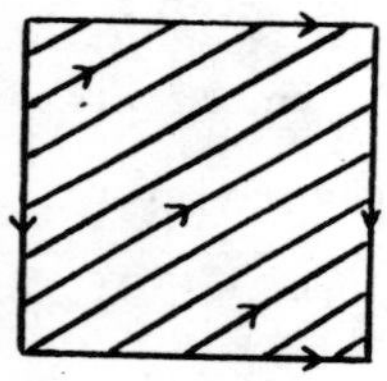

g),(1)

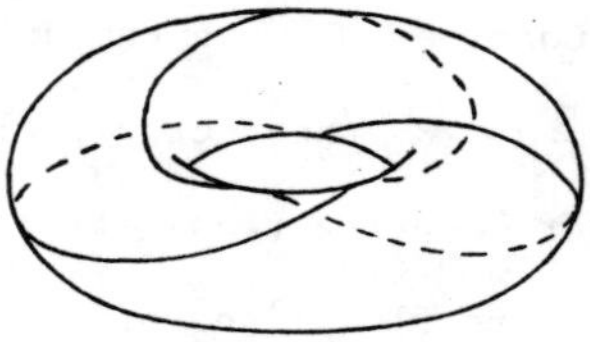

g),(2)

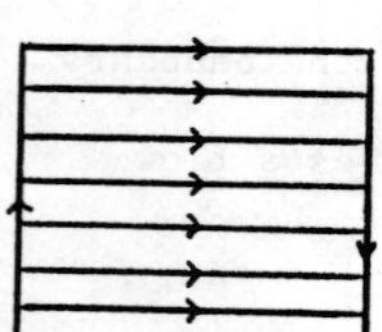

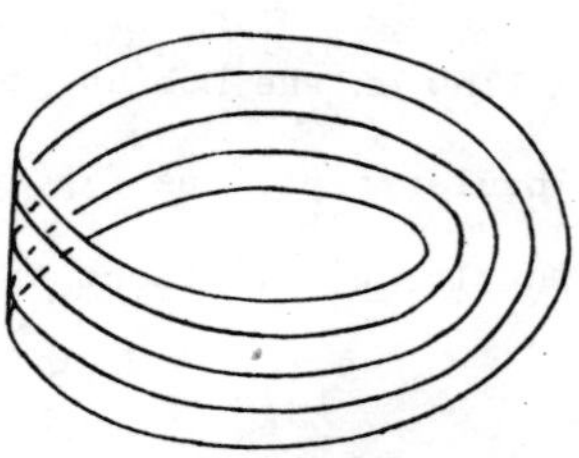

h)

i),(1)

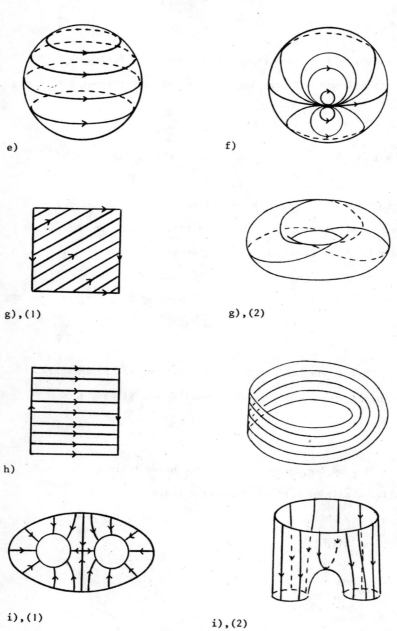

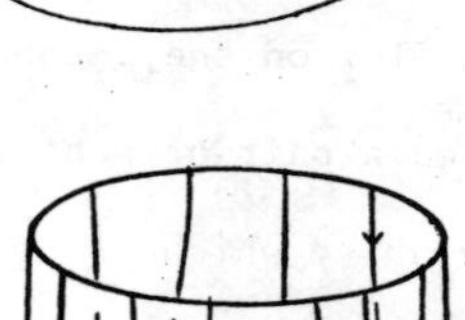

i),(2)

Convention. In future, whenever there is a flow on a manifold M with boundary it is supposed that on every boundary component M_o of M the flow is either tangent or transverse to M_o. In particular, no singularity will be contained in ∂M.

1.2. *The index of an isolated singularity*.

Let $x_o \in \mathbb{R}^2$ be the unique singularity of a C^1 vector field X which is defined in a disk neighbourhood D of x_o. Let c be an oriented simple closed piecewise differentiable curve in D encircling x_o. Then c bounds a disk $D_o \subset D$, by the Jordan-Schönflies theorem.

For each $x \in c$ let $\alpha(x) \in [0,2\pi)$ be the oriented angle between the vectors $e_1 = (1,0)$ and $X(x)$, cf. figure 3. If x runs on c in positive direction until it returns to its initial position then the angle $\alpha(x)$ has changed by a multiple of 2π. It is not hard to see that this number does not depend on the choice of the loop c and is invariant under orientation preserving diffeomorphism of D (see Milnor [Mi] for the precise statement). We denote it by $I(X,x_o)$ and call it the <u>index</u> of X at x_o.

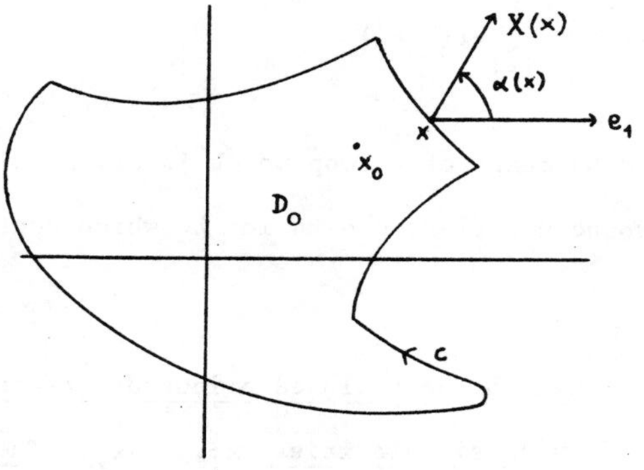

Figure 3

We may think of α as a map of c onto S^1. Then $I(X,x_o)$ is just the degree of α. If the orientation of c is changed then $I(X,x_o)$ changes its sign.

For regular points $x \in D$ of X we put $I(X,x) = 0$.

For a vector field X with isolated singularities on an oriented surface Σ, the index is well defined ; for each singularity x_i, take a positively oriented disk containing x_i and compute $I(X,x_i)$ by means of a loop contained in D_i.

1.2.1. - Exercise. Using a path encircling the origin in the counterclockwise direction show that $I(X,0) = 1$ for examples 1.1., a)-c) and $I(X,0) = -1$ for 1.1., d).

1.3. The theorem of Poincaré-Bohl-Hopf.

A general version of the following theorem of Poincaré-Bohl is due to H. Hopf; see Milnor [Mi].

Let $x_1,\dots,x_r$ be the singularities of the vector field X defined on the closed oriented surface Σ. Then we set

$$I(X) = \sum_{i=1}^{r} I(X,x_i)$$

where $I(X,x_i)$ is computed by means of a loop which is oriented as the boundary of the corresponding oriented disk in Σ which contains x_i.

1.3.1. - Theorem.- Let Σ be a closed oriented surface and X a vector field on Σ with singularities $x_1,\dots,x_r$. Then

$$\chi(\Sigma) = I(X) = \sum_{i=1}^{r} I(X,x_i)$$

where $\chi(\Sigma)$ is the Euler characteristic of Σ.

Proof : First we show $I(X) = I(Y)$ for any other vector field Y on Σ with only isolated singularities.

To this end we choose a triangulation T of Σ such that :

(1) The singularities of X and Y all lie in the complement of the 1-skeleton of T.

(2) Each face of T contains at most one singularity of X and one of Y.

Let $\Delta_j \in T$ be a face and c_j its boundary provided with the orientation induced from Δ_j. (The Δ_j are coherently oriented by the orientation of Σ). We choose two points x_j, y_j in $\mathring{\Delta}_j$ which should be singularities of X or Y if there are any in Δ_j.

For $x \in c_j$ let $\alpha(x)$ be the oriented angle between $X(x)$ and $Y(x)$, cf. fig. 4.

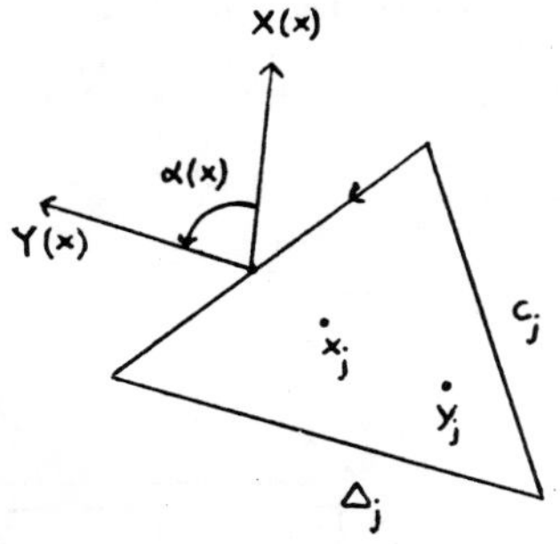

Figure 4

This defines a map

$$\alpha : c_j \to S^1$$

with degree equal to $I(Y,y_j) - I(X,x_j)$. Summing over all triangles of T gives

$$I(Y) - I(X) = \sum_j (I(Y,y_j) - I(X,x_j)).$$

But every edge in T appears exactly twice in the boundary of the faces of T, and with opposite orientations. It follows that the right hand side of the equation vanishes and thus $I(X) = I(Y)$.

It remains to construct a vector field X on Σ such that $I(X)$ equals the Euler characteristic of Σ.

On S^2 and T^2 we have already seen that there are such vector fields, cf. 1.2.1.

Let us assume that the genus g of Σ is greater than one. Recall the vector field X_o on the 2-sphere with three holes, Σ_o, as given in 1.1., i). To obtain a suitable vector field X on Σ we take $g-1$ copies of (Σ_o, X_o) and $g-1$ copies of $(-\Sigma_o, -X_o)$ and glue them together alternately as indicated in figure 5. The two free ends have to be identified. Since $I(X_o) = -1$ we get

$$I(X) = (-1)2(g-1) = \chi(\Sigma).$$

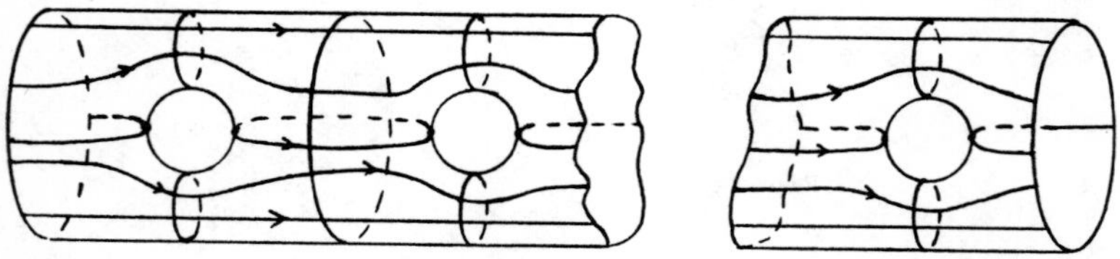

Figure 5

1.4. *Existence of non-singular vector fields.*

Vector fields without singularities yield foliations, as we shall see soon. Therefore it is worth stating the next theorem which is a corollary of 1.3.1.

1.4.1. - Theorem. Let Σ be a compact surface admitting a vector field without singularities.

i) If Σ is orientable then $\Sigma = T^2$ or $S^1 \times I$.

ii) If Σ is non-orientable then $\Sigma = K^2$ or $S^1 \times_{\mathbb{Z}_2} I$ (Möbius band).

Proof: If Σ is closed and orientable, then $\Sigma = T^2$, by 1.3.1. Now let Σ be orientable, of genus g and with $r \geq 1$ boundary components. We glue disks along the boundary components of Σ to get a closed surface Σ'.

A vector field on Σ without singularities can be extended to a vector field on Σ' with r singularities of index 1. By 1.3.1., it follows $2(1-g) = r \geq 1$. Therefore $g = 0$ and $r = 2$, that is $\Sigma = S^1 \times I$.

Clearly, $S^1 \times I$ admits a vector field without singularities.

If Σ is non-orientable then any vector field X on Σ can be lifted to a vector field $\tilde{X}$ on the orientation covering $\tilde{\Sigma}$ of Σ. If X has no singularities then $\tilde{X}$ is also without singularities. It follows that Σ must be either the Klein bottle or the Möbius band. Vector fields without singularities on these two surfaces have already been given in 1.1. h). □

2. *Foliations on surfaces.*

2.1. *Motivating remarks.*

Let X be a vector field without singularities on the surface Σ. According to the existence and uniqueness theorem on the solutions of ordinary differential equations the flow lines of X form a (locally trivial) family of curves which covers Σ. By contrast, let us consider the following situations.

a) The Klein bottle K^2 is (the total space of) an S^1-bundle over S^1. Clearly, the fibres of this bundle cover K^2 in a locally trivial way but there is no vector field on K^2 tangent to the fibres.

b) Take any vector field on K^2 and take the associated normal bundle. Then this normal bundle is integrable but without tangent vector field.

c) On $\mathbb{R}^2$, take a vector field X as indicated in fig. 6. The family of integral curves of X (but not the vector field itself) is invariant under the covering translations of $\pi : \mathbb{R}^2 \to T^2 = \mathbb{R}^2/\mathbb{Z}^2$. Thus we get a family of curves on T^2 without tangent vector field.

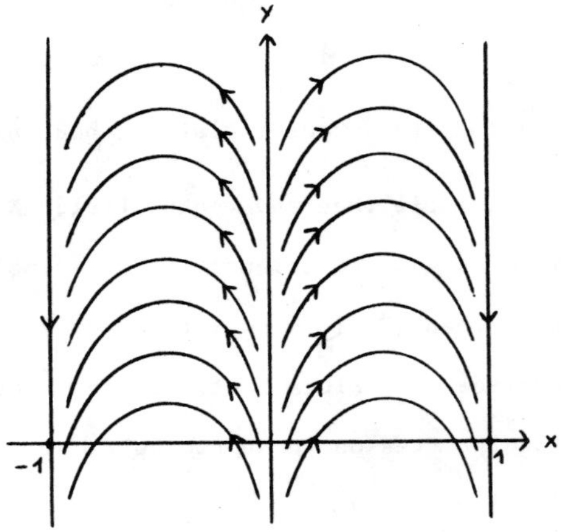

Figure 6

All these examples have in common

(1) a family $\{L_\alpha\}$ of curves which covers the surface Σ and which may be

(2) locally defined by a vector field without singularities.

Condition (2) may be replaced by :

(2') For every $p \in \Sigma$ there is a neighbourhood U of p in Σ and a homeomorphism

$$\varphi : U \to \varphi(U) \subset \mathbb{R}^2 = \mathbb{R} \times \mathbb{R}$$

such that $\varphi^{-1}(\mathbb{R} \times \{y\} \cap \varphi(U))$ lies in some L_α, for every $y \in \mathbb{R}$.

This motivates the following definition which also applies in the C^0 case.

2.2. *Definition of foliations and related notions.*

2.2.1. - *Definition.* Let Σ be a connected surface.

i) A foliation on Σ consists of a maximal atlas $\{(U_i, \varphi_i)\}$ of Σ such that the transition maps

$$\varphi_{ij} = \varphi_i \circ \varphi_j^{-1} | \varphi_j(U_i \cap U_j) : \varphi_j(U_i \cap U_j) \to \varphi_i(U_i \cap U_j)$$

are of the form

$$\varphi_{ij}(x,y) = (\alpha_{ij}(x,y), \gamma_{ij}(y)).$$

ii) If $\partial\Sigma \neq \emptyset$ then i) has to be modified so that for each component Σ_o of $\partial\Sigma$ the foliation becomes either tangent or transverse to Σ_o. For points $p \in \Sigma_o$ this means that there is a chart $\varphi_i : U_i \to \mathbb{H}^2$ around p where $\mathbb{H}^2$ is either $\mathbb{R} \times \mathbb{R}_+$ or $\mathbb{R}_+ \times \mathbb{R}$, respectively. Fig. 7 indicates the situation.

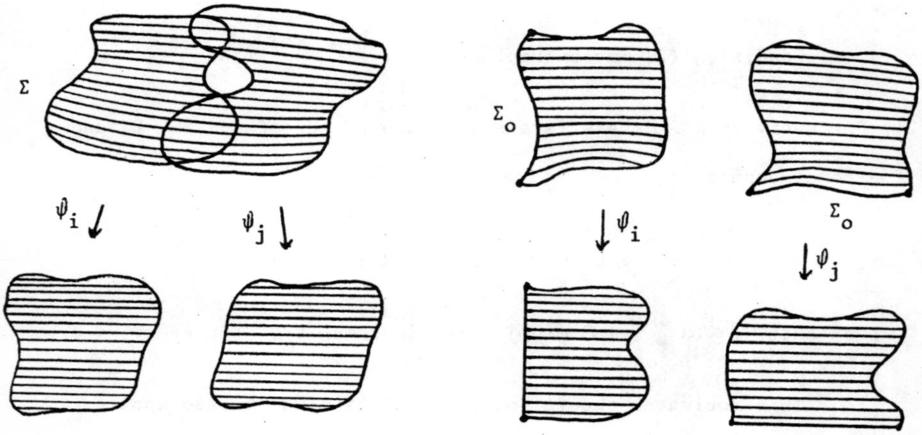

Figure 7

iii) A foliation is <u>of class</u> C^r if all (U_i,φ_i) are C^r charts, $0 \leqslant r \leqslant \infty$.

2.2.2. - <u>Notations</u>.- Usually, a foliation on the surface Σ will be denoted by (Σ,F) or simply by F, F_i, $\tilde{F}$, etc. We use also <u>foliated surface</u> as terminology.

The atlas $\{(U_i,\varphi_i)\}$ used in the definition of F is called a <u>foliated atlas</u>. The charts (U_i,φ_i) are referred to as <u>distinguished charts</u>, the U_i as <u>distinguished open sets</u> or <u>distinguished open (closed) squares</u> if $\varphi_i(U_i) = (-1,1) \times (-1,1)$ (or $\varphi_i(U_i) = [-1,1] \times [-1,1]$, respectively), and similarly when $\partial\Sigma \neq \emptyset$. It is clear that every $p \in \Sigma$ is contained in a distinguished open square.

Let $pr : \mathbb{R}^2 \to \mathbb{R}$ be the projection $pr(x,y) = y$. A map $f = pr \circ \varphi : U \to \mathbb{R}$, with a distinguished chart (U,φ), is called a <u>distinguished map</u>.

Provide $\mathbb{R}^2 = \mathbb{R} \times \mathbb{R}$ with the topology $\mathcal{T}_o$ which is the product of the natural topology on the first factor and the discrete topology on the second factor. The (connected) components of $\mathbb{R}^2$ with respect to $\mathcal{T}_o$ are precisely the lines $y = \text{const}$. Let $\varphi : U \to \varphi(U)$ be a distinguished chart of $\mathcal{F}$. There exists a unique topology $\mathcal{T}_U$ on U such that $\varphi : (U,\mathcal{T}_U) \to (\varphi(U),\mathcal{T}_o)$ becomes a homeomorphism. The components of $(U,\mathcal{T}_U)$ are called <u>plaques</u>.

The family of plaques for the distinguished charts of $\mathcal{F}$ is the basis of a topology $\mathcal{T}_L$ on Σ which is called the <u>leaf topology</u>. The components of $\mathcal{T}_L$ are the <u>leaves</u> of $\mathcal{F}$. The leaf through the point $p \in \Sigma$ is usually denoted by L_p. We write $L \in \mathcal{F}$ to indicate that L is a leaf of $\mathcal{F}$.

Two foliations $(\Sigma,\mathcal{F})$ and $(\Sigma',\mathcal{F}')$ are called <u>homeomorphic</u> if there is a homeomorphism $\varphi : \Sigma \to \Sigma'$ which takes the leaves of $\mathcal{F}$ onto the leaves of $\mathcal{F}'$. We then also say that $(\Sigma,\mathcal{F})$ and $(\Sigma',\mathcal{F}')$, are <u>isomorphic</u> or <u>conjugate</u>.

2.2.3. - *<u>Convention</u>.* All surfaces appearing henceforth are supposed to have a countable basis.

The leaves of a foliated surface are what we imagine, but this needs a proof.

2.2.4. - *Proposition*. - Let (Σ, F) be a foliated surface. If $\partial\Sigma = \emptyset$ then the leaves of F are homeomorphic to either S^1 or $\mathbb{R}$. If $\partial\Sigma \neq \emptyset$ then the leaves of F are homeomorphic to one of S^1, $\mathbb{R}$, I, $\mathbb{R}^+$.

Proof : It is well known that a connected 1-dimensional manifold L is homeomorphic to one of S^1 or $\mathbb{R}$ or, if $\partial L \neq \emptyset$, to I or $\mathbb{R}^+$, provided it is Hausdorff and with a countable basis. Since Σ is Hausdorff we can immediately conclude that the leaves of F are Hausdorff. It remains therefore to show that every leaf has a countable basis.

Let $\mathcal{U} = \{U_i\}$ be a countable basis for Σ. We may suppose that all U_i are distinguished open sets. Let $P_1, \ldots, P_m$ be plaques, each P_i lying in some U_i. We call $(P_1, \ldots, P_m)$ an m-chain if $P_\mu \cap P_{\mu+1} \neq \emptyset$ for $1 \leq \mu \leq m-1$. Let L be a leaf and $p \in L$. By A_m we denote the set of all m-chains such that $p \in P_1$. It follows by induction on m that A_m is countable. The plaques of $\mathcal{U}$ lying in L form a basis for the topology of L. But this basis is countable since each of its elements lies in some A_m. □

As an immediate consequence we obtain :

2.2.5. - *Corollary*. - Every foliated surface contains uncountably many different leaves.

2.3. *Orientability ; relation with vector fields*.

Recall that a foliated surface (Σ, F) is defined by an atlas $\{(U_i, \varphi_i)\}$ where the transition maps are of the form

$$\varphi_{ij}(x,y) = (\alpha_{ij}(x,y), \gamma_{ij}(y)).$$

The γ_{ij} are homeomorphisms between open subsets of the y-axis and the φ_{ij} are homeomorphisms also with respect to the topology T_o on $\mathbb{R}^2$ as introduced in 2.2.2.

The lines $x = 0$ and $y = y_o$ all carry natural orientations. We can therefore decide whether γ_{ij} or φ_{ij}, regarded as homeomorphisms between 1-dimensional manifolds, preserve orientations or not.

2.3.1. *Definition*.- i) The foliation F is called transversely orientable if there is a subatlas of $\{(U_i,\varphi_i)\}$ such that all γ_{ij} coming from this subatlas are orientation preserving.

ii) The foliation F is called orientable if there is a subatlas of $\{(U_i,\varphi_i)\}$ such that all φ_{ij} coming from this subatlas are orientation preserving.

Note that the definitions of transversely orientable and orientable foliations also make sense on non-orientable surfaces.

A foliation F being orientable intuitively means that there exists a coherent orientation of all the leaves of F.

2.3.2. *Remark*. - The flow lines of a vector field without singularities on Σ give rise to an orientable foliation on Σ.

Conversely, every orientable foliation F of class C^1 on Σ admits a tangent vector field, i.e. there is a vector field on Σ whose flow lines are the leaves of F. This can be seen by first putting a riemannian metric on Σ and then assigning to each $p \in \Sigma$ a unit vector $X(p) \in T_pL_p \subset T_p\Sigma$ of the tangent space T_pL_p to the leaf L_p at p. Since F is orientable this can be done in a continuous manner and so defines a vector field X on Σ, as required. See also II, 2.3.

2.3.3. *Definition*. - Let $f : \Sigma^* \to \Sigma$ be a covering map. Every foliation F on Σ yields a foliation F^* on Σ^*. If F is defined by the atlas $\{(U_i,\psi_i)\}$, where, for simplicity, all U_i are supposed to trivialize f then F^* is defined by $\{(U_{i_j},\psi_{i_j})\}$ where the U_{i_j} are the components of $f^{-1}(U_i)$ and $\psi_{i_j} = \psi_i \circ f$.

The foliation F^* is said to be induced by f or the lift of F by f. It is usually denoted by f^*F.

To the foliated surface (Σ,F) are associated two 2-fold covering maps :

2.3.4. - *The transverse orientation covering.*

For each $p \in \Sigma$ we devide the set of all distinguished charts around p into two classes. Two such charts (U_i,ψ_i) and (U_j,ψ_j) belong to the same class if and only if the germ of the homeomorphism γ_{ij} at $\psi_j(p)$, where $\psi_{ij} = (\alpha_{ij},\gamma_{ij})$, is orientation preserving. According to Haefliger [Ha] each of these classes is called a germ of transverse orientation of F at p.

Now any distinguished chart (U,ψ) determines exactly one germ of transverse orientation at any point $p \in U$. The set of all these germs of transverse orientation is denoted by U^*. The sets U^* form a basis for a topology on Σ^* the set of germs of transverse orientation of F.

The map $\pi : \Sigma^* \to \Sigma$ which assigns to a germ of transverse orientation of F at $p \in \Sigma$ the point p itself is a 2-fold covering. It is called the transverse orientation covering of (Σ,F).

2.3.5. - *The tangent orientation covering* is obtained in a similar way as the transverse orientation covering. This time we put two distinguished charts (U_i,ψ_i) and (U_j,ψ_j) of (Σ,F) in the same class if and only if the transition map ψ_{ij} locally preserves the orientations of the 1-dimensional manifolds $\varphi_j(U_i \cap U_j)$ and $\varphi_i(U_i \cap U_j)$.

The 2-fold covering $\tau : \Sigma_* \to \Sigma$, obtained along the lines of 2.3.4., is referred to as the tangent orientation covering of F.

Attention (see exercises). The coverings π and τ associated to the foliated surface (Σ, F) should not be confused with the orientation covering of Σ.

2.3.6. - *Proposition*. - Let (Σ, F) be a foliated surface and let $\pi : \Sigma^* \to \Sigma$ and $\tau : \Sigma_* \to \Sigma$ be the transverse orientation covering and the tangent orientation covering of F, respectively. Then

i) $\pi^* F$ is transversely orientable, and F is transversely orientable if and only if π is trivial.

ii) $\tau^* F$ is orientable, and F is orientable if and only if τ is trivial.

Proof : The maximal atlas for $(\Sigma^*, \pi^* F)$ contains the subatlas of all distinguished charts of the form (U^*, φ^*), where (U, φ) is a distinguished chart of F, U^* is the set of germs of transverse orientation of F at the points of U, determined by (U, φ), and $\varphi^* = \varphi \circ (\pi | U^*)$.

Now suppose (U_i, φ_i) and (U_j, φ_j) are distinguished charts of F such that $U_i \cap U_j \neq \emptyset$ and such that for $p \in U_i \cap U_j$ the germ of the corresponding homeomorphism γ_{ij} at $\varphi_j(p)$ (cf. 2.2.1, i)) reverses the orientation of the y-axis. Then there exist neighbourhoods V_i and V_j of p, $V_i \subset U_i$, $V_j \subset U_j$, such that $V_i^* \cap V_j^* = \emptyset$, by the construction of π. It follows that $\pi^* F$ is transversely orientable.

If π is trivial then, by what was just proved, F is transversely orientable.

Conversely, if F is transversely orientable then there exists an atlas $\{(U_i,\varphi_i)\}$ of (Σ,F) such that all corresponding γ_{ij} are orientation preserving. Let $\Sigma_o^* \subset \Sigma^*$ be the union of all U_i^* then Σ_o^* is mapped by π homeomorphically onto Σ. We conclude that Σ^* consists of two components and thus π is trivial. This completes the proof of i).

The proof of ii) is analogous. □

We shall now introduce the concept of a transversal of a foliation. This concept will later turn out to be very useful.

2.3.7. - *Definition*. - Let (Σ,F) be a foliated surface.

i) A simple curve c of the interval J in Σ is called transverse to F in the point $p = c(t)$ if there is a distinguished map (U,f) around p such that $f \circ c$ is injective in a neighbourhood of $t \in J$. (This condition does not depend on the choice of (U,f)).

The curve c is called a transversal of F if c is transverse to F in each of its points.

ii) A foliation $F^{\pitchfork}$ on Σ is transverse to F if every leaf of $F^{\pitchfork}$ is a transversal of F.

2.3.8. - *Proposition*. - Every foliation (Σ,F) admits a transverse foliation $(\Sigma,F^{\pitchfork})$. Moreover, if F is of class C^r then there exists $F^{\pitchfork}$ of class C^r.

For $r > 0$ and F transversely orientable this follows by an argument similar to that in 2.3.2 using a normal vector field instead of a tangent vector field to F. For arbitrary differentiable F we refer the reader to II, 2.3.

The proposition also holds for foliations of class C^0. A proof will be given in 2.4.7.

2.3.9. *Definition.* - Let $F^{⋔}$ be a transverse foliation of (Σ, F) and let $s : \mathbb{R}^2 \to \mathbb{R}^2$ be defined by $s(x,y) = (y,x)$. A coordinate chart $\varphi : U \to \mathbb{R}^2$ of Σ is <u>bidistinguished with respect to</u> F <u>and</u> $F^{⋔}$ if (U,φ) and $(U, s \circ \varphi)$ are distinguished squares of F and $F^{⋔}$, respectively. That is to say φ takes the F-plaques to the horizontal intervals $(-1,1) \times \{y\}$ and the $F^{⋔}$-plaques to the vertical lines $\{x\} \times (-1,1)$. (Sometimes we shall permit U to be a closed square).

The definition applies in a similar way when $\partial\Sigma \neq \emptyset$.

Given $(\Sigma, F, F^{⋔})$ there exists of course a bidistinguished atlas (by taking all bidistinguished charts).

2.3.10. - <u>Lemma</u>. - <u>Let</u> (Σ, F) <u>be a foliated surface. Then the two following conditions are equivalent</u> :

(1) F <u>is transversely orientable</u>.

(2) <u>Every transverse foliation</u> $F^{⋔}$ <u>is orientable</u>.

Proof : Condition (2) clearly implies (1).

For the converse, let $A = \{(U_i, \varphi_i)\}$ be a maximal atlas which orients F transversely. Given $F^{⋔}$ there exist a subatlas $A^{⋔} \subset A$ which is bidistinguished for F and $F^{⋔}$ and so orients $F^{⋔}$. □

2.3.11. - *Proposition*. - <u>Let</u> F <u>be an orientable foliation on the surface</u> Σ. <u>We have two equivalent conditions</u> :

(1) Σ <u>is orientable</u>.

(2) F <u>is transversely orientable</u>.

Proof : Suppose Σ is orientable and take a maximal atlas A which orients Σ. There exists a subatlas $A' \subset A$ which defines F and orients F. Then clearly A' orients F transversely.

The converse statement is proved in the same way. □

2.3.12. - *Exercises*. i) Which of the foliations indicated or given explicitly in the previous sections are (transversely) orientable?

ii) Give an example of a foliation on the Klein bottle which is neither orientable nor transversely orientable.

iii) Describe the relationship between the transverse orientation covering, the tangent orientation covering of a foliated surface (Σ,F) and the orientation covering of Σ.

2.4. *The existence theorem of Poincaré-Kneser*.

It is known that every non-compact surface admits vector fields without singularities and therefore foliations. We shall now specify all compact surfaces which admit foliations. The surfaces admitting foliations of class C^1 have already been determined by 1.4.1, 2.3.2 and 2.3.6, ii). But thus far nothing was said with regard to C^0 foliations.

The technical tool which will be used in this section is a "nice" triangulation of any foliated surface. Some preliminaries are necessary to construct this triangulation.

2.4.1. *Definition*. - Let (Σ,F) be a foliated surface.

i) A curve $c : I \to \Sigma$ is said to be in <u>general position</u> (<u>with respect to</u> F) if the set of points $p \in c(I)$ such that c is not transverse to F in p is finite. These points are then called <u>contact points</u>.

ii) A triangulation T of Σ is in general position (with respect to F) if

(1) every face of T is contained in a distinguished square,

(2) the edges of T are in general position,

(3) the vertices of T and the contact points all belong to different leaves.

iii) A triangulation T of Σ is transverse to F if T is in general position and every edge of T is transverse to F.

2.4.2. - *Lemma*. - Let F_o be the "horizontal" foliation and F_1 the "vertical" foliation on $\mathbb{R}^2$, let $c : I \to \mathbb{R}^2$ be a simple curve such that $c(0)$ and $c(1)$ lie on different leaves of F_o and of F_1 and let U be an open neighbourhood of $c(I)$. Then there exists a simple curve $c' : I \to \mathbb{R}^2$ such that

(1) $c'(0) = c(0)$, $c'(1) = c(1)$ and $c'(I) \subset U$.

(2) c' is in general position with respect to F_o and to F_1.

Proof : Take a polygonal approximation c' of c modulo its endpoints. This clearly can be found in U, in such a way that all of its vertices lie on different leaves of F_o and of F_1. Thus c' is in general position with respect to F_o and F_1. □

2.4.3. - *Proposition*. - Let (Σ,F) be a compact foliated surface then there exists a triangulation of Σ which is in general position with respect to F.

Proof : Given any triangulation of Σ we subdivide it so that every face of the so obtained triangulation T lies in a distinguished square.

Let $v_1,\dots,v_s$ be the vertices of $\mathcal{T}$. For each v_j we choose a distinguished closed square S_j containing v_j and so that $S_i \cap S_j = \emptyset$ for $i \neq j$. Let e be an edge of $\mathcal{T}$, with endpoints v_i and v_j. Among the components of $e - (\mathring{S}_i \cup \mathring{S}_j)$ there is one, we denote it by $\tilde{e}$, that joins ∂S_i and ∂S_j. Now for each edge e_k we choose a neighbourhood U_k of $\tilde{e}_k$ in Σ such that $U_k \cap U_\ell \neq \emptyset$ implies $e_k = e_\ell$. Using 2.4.2. we approximate each $\tilde{e}_k$ within U_k by a curve e'_k in general position with respect to $\mathcal{F}$. We may moreover suppose that $e'_k \cap (\cup \mathring{S}_j) = \emptyset$. We connect all the endpoints of different edges e'_k on ∂S_j to v_j by affine arcs. The so extended edge e'_k is denoted by $\hat{e}_k$. The situation is indicated in figure 8.

As there are uncountably many leaves we may assume, possibly after a small isotopy of Σ, that all vertices and contact points involved in the above constructions lie on different leaves of $\mathcal{F}$.

Now if Δ is a face of $\mathcal{T}$ with edges e_1, e_2, e_3 the curves $\hat{e}_1, \hat{e}_2, \hat{e}_3$ define a Jordan curve in some distinguished square and so bound a 2-cell in Σ. This shows that the $\hat{e}_k$ determine a triangulation of Σ with the required properties. □

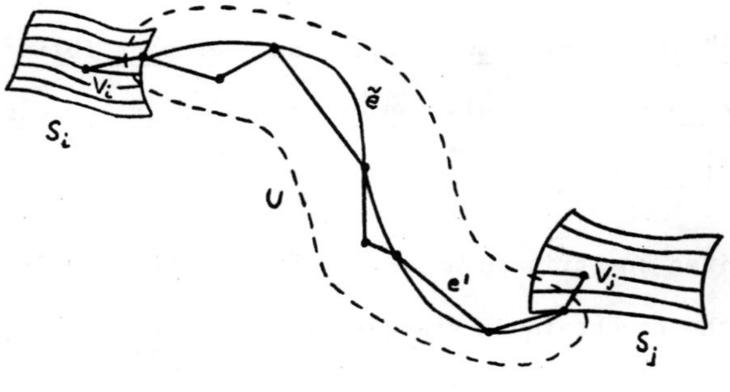

Figure 8

2.4.4. - Proposition. - Let (Σ, F) be a compact foliated surface. There exists a triangulation of Σ transverse to F.

Proof : Let T be a triangulation of Σ in general position with respect to F and let Δ be a face of T. We denote by Δ' the image of Δ under the distinguished chart $\varphi : U \to \mathbb{R}^2$ with $\Delta \subset U$. Then Δ' is in general position with respect to the horizontal foliation $(\mathbb{R}^2, F_o)$. It suffices to show that we can subdivide Δ' in Δ'' without introducing new vertices and such that Δ'' is transverse to F_o.

Let $P = \{p_1, \ldots, p_s\} \subset \mathbb{R}^2$ be the set consisting of the three vertices of Δ' and the contact points of $\partial\Delta'$, numbered in such a way that

$$pr(p_1) < \ldots < pr(p_s) ,$$

where $pr : \mathbb{R}^2 \to \mathbb{R}^2$ is the projection $(x,y) \mapsto y$.

The point p_1 is an endpoint of two transverse edges e_1 and e_2 of $\partial\Delta'$ with second endpoint p_{i_1} and p_{i_2}, respectively. We may suppose $p_{i_1} < p_{i_2}$. Let h be the horizontal path joining p_{i_1} and $p \in e_2$ and let e be the path on e_2 joining p and p_1. The loop $J = e_1 * h * e$ defines a Jordan curve which bounds a 2-cell C in $\mathbb{R}^2$ (see fig. 9).

There are two possibilities :

a) if $p_2 \notin C$ then we can approximate the curve $h * (e_2 - e)$ modulo boundary by a transverse curve k such that $k \cap \partial\Delta' = \{p_{i_1}, p_{i_2}\}$,

b) if $p_2 \in C$ then we can join p_1 and p_2 by a transverse curve k such that $k \cap \partial\Delta' = \{p_1, p_2\}$.

In both cases k subdivides Δ' into two triangles Δ_1 and Δ_2. If P_i denotes the set of vertices and contact points of $\partial\Delta_i$ then the cardinality of P_i is smaller than that of P. Thus, after finitely many steps we get Δ'' as required. □

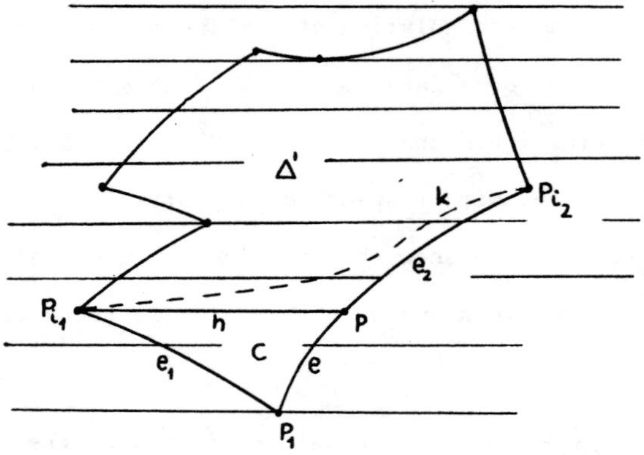

Figure 9

2.4.5. *Proposition*. - Let Σ be closed, T a triangulation transverse to the foliation (Σ, F) and A_o and A_2 the numbers of vertices and faces of T, respectively. Then $2A_o = A_2$.

Proof : To every vertex v of T there are exactly two faces, $\Delta_1 \subset U_1$ and $\Delta_2 \subset U_2$, of T which contain v and such that the plaques of the distinguished open sets U_1 and U_2 containing v have non-empty intersection with $\mathring{\Delta}_1$ and $\mathring{\Delta}_2$. This provides a two-one correspondence between the faces and the vertices of T. □

We are now able to decide which compact surfaces admit C^o foliations.

2.4.6. - *Theorem* (*Poincaré-Kneser*). - Let Σ be a compact surface and $0 \leqslant r \leqslant \infty$. Then Σ admits a foliation of class C^r

<u>if and only if its Euler characteristic is zero</u>, <u>i.e.</u> Σ <u>is a torus</u>, <u>Klein bottle</u>, <u>annulus or Möbius band</u>.

Proof : If $\chi(\Sigma) = 0$ we have already seen (and shall again see later) that there exist (orientable) C^r foliations on Σ, for all r.

For $r \geq 1$ the theorem follows from 1.3. and 2.3.6, ii) since every orientable C^1 foliation F on Σ gives rise to a vector field on Σ whose orbits are the leaves of F. It therefore remains to consider the case $r = 0$.

Let Σ be closed and assume that there is a foliation on Σ. We choose a triangulation T of Σ, as in 2.4.4. By A_o, A_1, A_2 we denote the number of vertices, edges and faces of T, respectively. Then obviously

$$3A_2 = 2A_1.$$

On the other hand

$$2A_o = A_2 ,$$

by 2.4.5. It therefore follows

$$\chi(\Sigma) = A_o - A_1 + A_2 = A_o - 3A_o + 2A_o = 0.$$

(We found this ingenious argument in Kneser [Kn]).

The case $\partial\Sigma \neq \emptyset$ is reduced to the case $\partial\Sigma = \emptyset$. If $\partial\Sigma \neq \emptyset$ and F is a foliation on Σ then by identifying antipodal points on the boundary components of Σ we get a foliation on a closed non-orientable surface which must be the Klein bottle. We deduce that Σ must have been either the annulus or the Möbius band. □

At the end of 2.3.8. we promised to show the existence of a transverse foliation also in the C^o case. This will be done now.

2.4.7. - *Proof of 2.3.8. for foliations of class C^o on compact surfaces.*

Let T be a triangulation of (Σ, F), as in 2.4.4. To every vertex v of T there is a distinguished chart (U,φ) centered at v such that the image of the part of T in U under φ looks like the cone in $0 \in \mathbb{R}^2$ over a convex polygon whose vertices do not lie on the y-axis, see fig. 10.

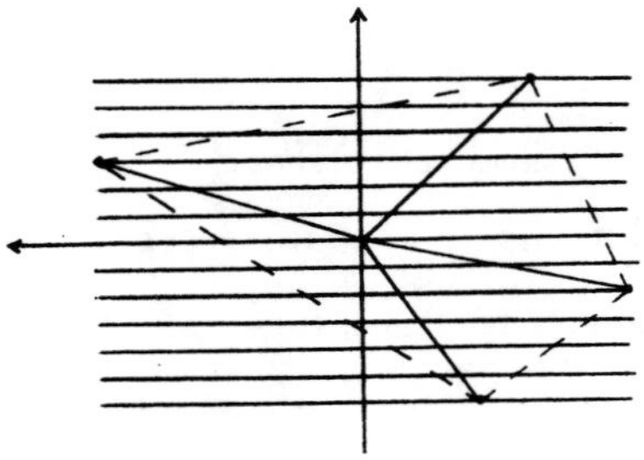

Figure 10

In a small neighbourhood V of 0 in $\varphi(U)$ we choose as transverse foliation the vertical foliation, see fig. 10. This is done for every vertex of T and so defines $F^{\pitchfork}$ in a neighbourhood of the vertices of T.

We now extend this foliation over the 1-skeleton of T. This can be done in such a way that the transverse foliation is also transverse to each edge of T, with the exception of at most one point, depending on the transverse foliation already given in neighbourhoods of its endpoints, see fig. 12.

Consider the image Δ of a face of T under a distinguished chart of F. As the edges of Δ are transverse to the horizontal foliation F_o it follows that two of the vertices of Δ, denote them by v_1 and v_2, are contact points of $\partial\Delta$ with respect to F_o. We have to distinguish three cases for v_1, and similarly for v_2, and three cases for the third vertex v_3 of Δ. These are indicated in fig. 11.

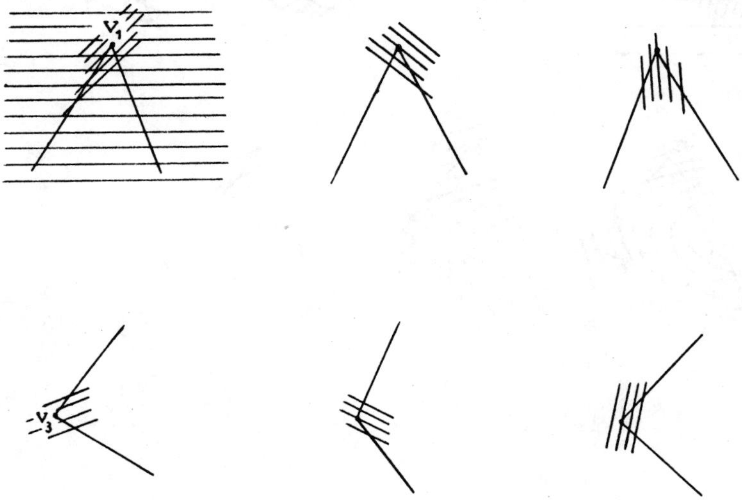

Figure 11

The transverse foliation is indicated in fig. 11 by line segments transverse to F_o. There are twenty seven possibilities how the transverse foliation in a neighbourhood of $\partial\Delta$ can look. These are all essentially of one of the seven types listed in fig. 12.

It is easily seen that in all cases the transverse foliation that is defined in a neighbourhood of $\partial\Delta$ can be transversely extended to the interior of Δ, see fig. 12. □

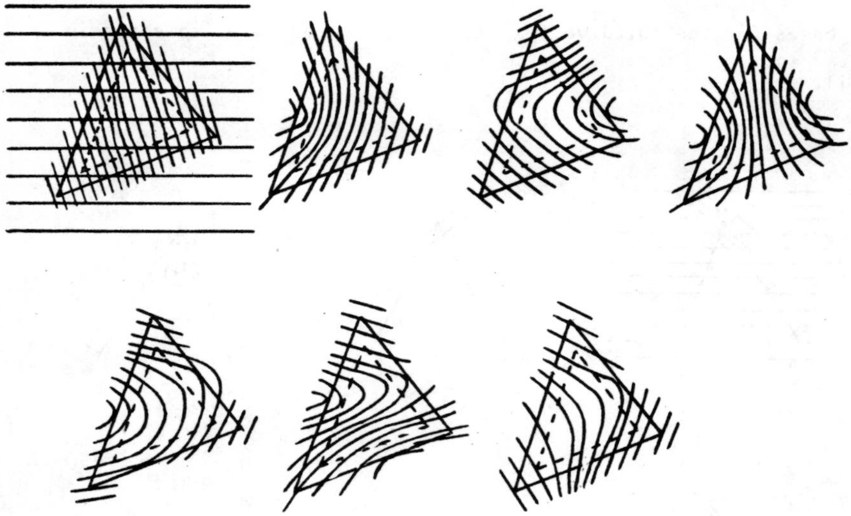

Figure 12

2.4.8. - *Exercises.* i) Adapt proposition 2.4.4. to the case of foliated surfaces with boundary. Moreover, if the foliation F on $(\Sigma,\partial\Sigma)$ is transverse to $\partial\Sigma$ show that there exists $F^{\pitchfork}$ tangent to $\partial\Sigma$.

ii) (Construction of $F^{\pitchfork}$ for open foliated surfaces). Let (Σ,F) be an open foliated surface.

a) Show that every compact surface Σ_o in Σ can be approximated by a compact surface Σ_1 such that $\partial\Sigma_1$ is in general position with respect to F.

b) Let now $\Sigma_o \subset \Sigma$ be a compact surface such that $\partial\Sigma_o$ is in general position. Construct a triangulation T_o of Σ_o, transverse to F, and show that there exists an open neighbourhood U_o of Σ_o and a foliation $F_o^{\pitchfork}$ of U_o such that T_o and $F|U_o$ are transverse to $F_o^{\pitchfork}$.

c) Let $\Sigma_1 \subset \Sigma$ be a compact surface such that $\Sigma_o \subset U_o \subset \Sigma_1$ and $\partial\Sigma_1$ is in general position with respect to F. Show that, after a suitable subdivision, T_o can be extended to a triangulation T_1 of Σ_1 transverse to F and to $F_o^{\pitchfork}$. Deduce that there exists an open neighbourhood U_1 of Σ_1 such that $F_o^{\pitchfork}$ can be extended to U_1, transversely to F and so that T_1 is transverse to the extended foliation.

d) Construct a transverse foliation $F^{\pitchfork}$ for any foliation (Σ, F).

3. *Construction and description of foliations.*

3.1. *Suspensions.*

3.1.1. - *Definition and notation.* - For $0 \leq r \leq \infty$ we denote by $\mathrm{Diff}^r(M)$ the group of C^r diffeomorphisms of the manifold M (C^o diffeomorphisms are homeomorphisms, the group $\mathrm{Diff}^o(M)$ is also denoted by Homeo(M)).

If M is oriented then $\mathrm{Diff}_+^r(M) \subset \mathrm{Diff}^r(M)$ is the group of orientation preserving diffeomorphisms.

The manifolds $S^1 = \mathbb{R}/\mathbb{Z}$, $I = [0,1]$, $\mathbb{R}$, $\mathbb{R}^+ = \{x \in \mathbb{R} \mid x \geq 0\}$ are all canonically oriented. (S^1 gets the orientation such that the quotient map

$$q : \mathbb{R} \to S^1 = \mathbb{R}/\mathbb{Z}$$

is orientation preserving).

The groups $\mathrm{Diff}_+^r(S^1)$ will play an important role in the sequel. We may think of them as obtained from $\mathrm{Diff}_+^r(\mathbb{R})$ in the following way.

Let $D^r(S^1)$ be the group consisting of all $f \in \mathrm{Diff}^r_+(\mathbb{R})$ such that $\psi = f - \mathrm{id}_{\mathbb{R}}$ is $\mathbb{Z}$-periodic, i.e. $\psi(x+1) = \psi(x)$ for all $x \in \mathbb{R}$.

If $R_\alpha : \mathbb{R} \to \mathbb{R}$, $\alpha \in \mathbb{R}$, is the translation, or shift, by α, i.e. $R_\alpha(x) = x+\alpha$, then the center C of $D^r(S^1)$ consists of all R_k, $k \in \mathbb{Z}$.

Finally,

$$\mathrm{Diff}^r_+(S^1) = D^r(S^1)/C.$$

For $f \in D^r(S^1)$ we denote by $\bar{f}$ the corresponding element in $\mathrm{Diff}^r_+(S^1)$. The element of $\mathrm{Diff}^r_+(S^1)$ of the form $\bar{R}_\alpha$, $\alpha \in \mathbb{R}/\mathbb{Z}$, is called the rotation of S^1 through the angle α (mod 1).

3.1.2. - *Suspension*. - Let M be one of S^1, $\mathbb{R}$, $\mathbb{R}^+$ or I and let

$$\chi : \pi_1 S^1 \to \mathrm{Diff}^r(M)$$

be a representation. Then there is a free action of $\mathbb{Z}$ on $M \times \mathbb{R}$, namely $\mathbb{Z}$ acts on $\mathbb{R}$ as group of covering translations of $q : \mathbb{R} \to S^1$ and on M by χ. So for θ a generator of $\pi_1 S^1$ and $f = \chi(\theta)$ we have

$$k(t,x) = (f^k(t), x + k), \ x \in \mathbb{R},$$

where $k \in \mathbb{Z}$ and $f^k = f \circ \ldots \circ f$ (k times).

We get the following commutative diagram

$$\begin{array}{ccc} M \times \mathbb{R} & \xrightarrow{\ pr\ } & \mathbb{R} \\ \downarrow \pi & & \downarrow q \\ \Sigma_f = (M \times \mathbb{R})/\mathbb{Z} & \dashrightarrow^{p} & S^1 \end{array}$$

where pr is the projection, π is the quotient map and p is canonically induced.

Evidently,

$$\Sigma_f = \begin{cases} T^2 & \text{if } M = S^1 \text{ and } f \text{ is orientation preserving} \\ K^2 & \text{if } M = S^1 \text{ and } f \text{ is orientation reversing} \\ S^1 \times I & \text{if } M = I \text{ and } f \text{ is orientation preserving} \\ S^1 \times_{\mathbb{Z}_2} I & \text{if } M = I \text{ and } f \text{ is orientation reversing} \\ S^1 \times \mathbb{R} & \text{if } M = \mathbb{R} \text{ and } f \text{ is orientation preserving} \\ S^1 \times_{\mathbb{Z}_2} \mathbb{R} & \text{if } M = \mathbb{R} \text{ and } f \text{ is orientation reversing} \\ S^1 \times \mathbb{R}^+ & \text{if } M = \mathbb{R}^+. \end{cases}$$

On $M \times \mathbb{R}$ we have the vertical foliation $\mathcal{F}_o$ by lines $\{t\} \times \mathbb{R}$, $t \in M$. This foliation is preserved by the action of $\mathbb{Z}$, i.e. for every leaf L of $\mathcal{F}_o$ and every $k \in \mathbb{Z}$ the set $k(L)$ is again a leaf of $\mathcal{F}_o$. Since the action of $\mathbb{Z}$ on $M \times \mathbb{R}$ is free, properly discontinuous and of class C^r the quotient map $\pi : M \times \mathbb{R} \to \Sigma_f$ is a regular covering and we get a C^r foliation $\mathcal{F}_f$ on Σ_f, according to the following lemma. The foliation $(\Sigma_f, \mathcal{F}_f)$ is called the suspension of the diffeomorphism f of M.

3.1.3. - *Lemma.* - Let $(\Sigma_o, \mathcal{F}_o)$ be a foliated surface and $\pi : \Sigma_o \to \Sigma$ a regular covering with group of covering translations Γ. If $\mathcal{F}_o$ is preserved by the action of Γ then there exists a foliation $\mathcal{F}$ on Σ such that $\mathcal{F}_o = \pi^* \mathcal{F}$.

Moreover, if $\mathcal{F}_o$ and π are of class C^r then $\mathcal{F}$ is of class C^r.

Proof : For every $x \in \Sigma$, there exists an open neighbourhood U_x of x and a point $y \in \pi^{-1}(x)$ such that

(1) U_x trivializes π,

(2) the component U_y of y in $\pi^{-1}(U_x)$ is the domain of a distinguished chart (U_y, φ_y) of F_o.

As F_o is preserved by Γ it follows that $(\gamma(U_y), \varphi_y \circ \gamma^{-1})$ is also a distinguished chart of F_o, for each $\gamma \in \Gamma$. Now if $U_{x_1} \cap U_{x_2} \neq \emptyset$. then there exists a $\gamma \in \Gamma$ such that $U = U_{y_1} \cap \gamma(U_{y_2}) \neq \emptyset$ and we have the commutative diagram

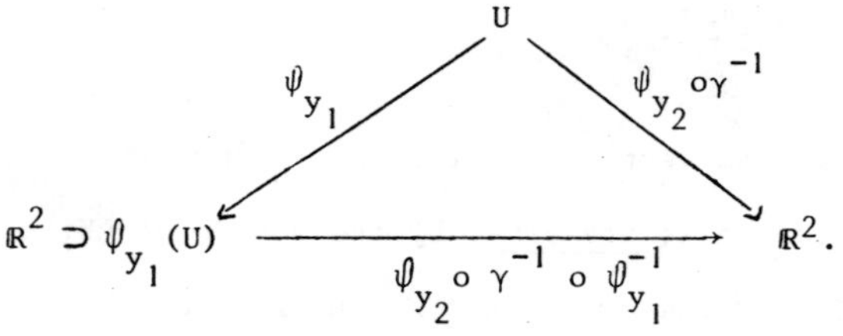

Thus, if $S_y : U_x \to U_y$ is the local section of π then the atlas $\{(U_x, \varphi_y \circ S_y)\}$ defines a foliation F on Σ which is of the same differentiability class as F_o. By construction, we have $F_o = \pi^* F$. □

Example. A special case of suspensions on T^2 are the <u>linear foliations</u>. These are obtained by rotations of S^1. The leaves of the suspension F wind around T^2 with constant slope. If the angle α is irrational then all leaves are real lines and dense in T^2. If $\alpha \in \mathbb{Q}$ then all leaves are circles and the foliation is a product. see also 1.1.,g).

Here are two properties of suspensions :

(1) For every $L \in F$ the map $p|L : L \to S^1$ is a covering map.

(2) The compact leaves of F are in one-one correspondence with the orbits of periodic points of f. (The point $t \in M$ is periodic if $f^k(t) = t$ for some $k \in \mathbb{N}$).

3.1.4. - *Exercises*. i) Let T^2 be endowed with a linear foliation F.

a) Show that there exists a closed 1-form ω on T^2 with real coefficients whose kernel is a line field on T^2 which is tangent to F. (We say that ω defines F).

b) Show that, for every $\varepsilon > 0$, there exists a closed 1-form Ω on T^2 such that the norm $\|\omega - \Omega\| \leq \varepsilon$ and Ω defines a foliation which is a fibration of T^2 over S^1. (We say that F can be approximated by a fibration; compare also Tischler's theorem in chp.).

ii) Let M be one of $\mathbb{R}$, S^1, I, $\mathbb{R}^+$. If possible, give (non-trivial) examples of $f \in \mathrm{Diff}^r(M)$, orientation preserving or not, such that

a) f has a finite number of periodic points,

b) f has no periodic points, apart from the boundary points if $M = I$ or $\mathbb{R}^+$,

c) there is a Cantor set C invariant under f, i.e. $f(C) = C$. (A Cantor set in M, or in any 1-dimensional manifold homeomorphic to M, is a closed non-empty subset of M without isolated points and without interior points).

If possible draw pictures of the corresponding suspensions.

iii) Let M be as in ii). If $f,g \in \mathrm{Diff}^r(M)$ are C^r conjugate, i.e. if there is $h \in \mathrm{Diff}^r(M)$ such that $g = h \circ f \circ h^{-1}$, what can be said about the suspensions of f and g ?

3.2. *Germs near circle leaves ; leaf holonomy.*

3.2.1. *Definition*. - Let F be a foliation on Σ and let $\Sigma_o \subset \Sigma$ be an embedded surface, possibly with boundary. We say that F induces a foliation on Σ_o, denoted by $F|\Sigma_o$, when the family of the restrictions of the distinguished charts of F to Σ_o contains a foliated atlas of Σ_o. We then also say that $F|\Sigma_o$ is the restriction of F to Σ_o. Clearly, $F|U$ exists for every open set U of Σ. The leaves of $F|U$ are the components of the intersections of the leaves of F with U.

3.2.2. *Germs*. - Occasionally, we shall be interested in the behaviour of a foliation F in the neighbourhood of a circle leaf L.

The germ of F near L is by definition the family $\{F|U_j\}$ where $\{U_j\}$ is the family of open neighbourhoods of L. We take $g(F,L)$ as notation.

A representative of $g(F,L)$ is any restriction $F|U_j$. Two germs $g(F,L)$ and $g(F',L')$ are said to be homeomorphic if there is a homeomorphism between representatives of $g(F,L)$ and $g(F',L')$ taking L onto L'.

We now want to describe all germs $g(F,L)$ for L homeomorphic to $I = [0,1]$ or S^1.

3.2.3. *Lemma*. - Let F_o be the horizontal foliation of $I \times \mathbb{R}$ with leaves $I \times \{t\}$, $t \in \mathbb{R}$, and let I_o be the leaf $I \times \{0\}$. If the foliation (Σ, F) contains a leaf L homeomorphic to I then the germ $g(F,L)$ is homeomorphic to $g(F_o, I_o)$.

Proof : Our assumptions imply that $\partial\Sigma \neq \emptyset$ and that the foliation F is transverse to $\partial\Sigma$. By 2.3.8., in connection with 2.4.7., there is a transverse foliation $F^{\pitchfork}$ of F that may be supposed to be tangent to $\partial\Sigma$, (see 2.4.8., i)). We may find finitely many bidistinguished squares (U_j,φ_j), $j = 1,\ldots,s$, such that

(1) for every j, $U_j \cap L$ is the plaque defined by $y = 0$,

(2) $U_i \cap U_j$ is connected and $U_i \cap U_j \neq \emptyset$ implies $j = i + 1$,

(3) there are points $0 = t_o < \ldots < t_s = 1$ in $L \approx I$ such that $t_{j-1}, t_j \in U_j$.

After possibly shrinking U_1 and U_2 in the y-direction we may suppose that $U_1 \cup U_2$ is also a bidistinguished square. So we get a covering of L by s-1 bidistinguished squares and after finitely many repetitions of this process we get a single bidistinguished square covering L. This proves the lemma. □

With the same kind of argument as in the proof of the last lemma one proves the following more general result which may be considered as a kind of trivialization lemma.

3.2.4. - *Lemma*. - Let Σ be a surface with transverse foliations F and $F^{\pitchfork}$ and let c be a non-closed curve in a leaf of F. Then there is a bidistinguished (open or closed) square (U,φ) such that $c \subset \varphi^{-1}((-1,1) \times \{0\})$.

Of course, without a transverse foliation $F^{\pitchfork}$ this lemma could be stated for U beeing simply a distinguished square of F.

3.2.5. - *Proposition*. - Let $g(F,L)$ be the germ of a C^r foliation (Σ,F) near the circle leaf $L \subset \mathring{\Sigma}$. There exists a C^r diffeomorphism f of $\mathbb{R}$ with 0 as fixed point such that $g(F,L)$ is homeomorphic to $g(F_f,L_o)$ where L_o is the circle leaf of (Σ_f,F_f) corresponding to 0.

Proof : We first choose a transverse foliation $F^{\pitchfork}$. Then we take a bidistinguished square U_o such that $U_o \cap L$ is connected. Let x_1, x_o, x_2 be three points of $U_o \cap L$. We join x_1 and x_2 in L by a curve c that does not meet x_o and take a bidistinguished closed square (U_1, φ_1) containing c, according to 3.2.4. We may moreover assume that the two vertical boundary plaques J_1 and J_2 of U_1 belong to U_o, see fig. 13.

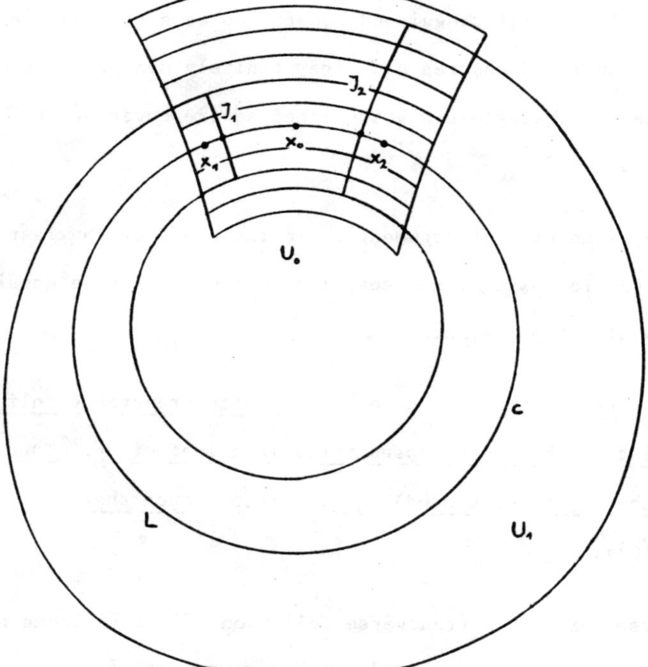

Figure 13

Now using $F|U_o$ we get a "local" C^r diffeomorphism $f_o : J_1 \to J_2$, i.e. a C^r diffeomorphism $f_o : J_1' \to J_2'$ between neighbourhoods of $J_i \cap L$ in J_i, $i = 1,2$, and taking $J_1 \cap L$ to $J_2 \cap L$. This shows that L has a closed neighbourhood V homeomorphic to

$U_1/x \sim f_o(x)$; see fig. 13. The C^r diffeomorphism $f_1 = \varphi_1 \circ f_o \circ \varphi_1^{-1}$ is defined in a neighbourhood of $0 \in \mathbb{R}$ and keeps 0 fixed. We may extend f_1 to a C^r diffeomorphism f of $(\mathbb{R},0)$. It then follows that $F|\mathring{V}$ is homeomorphic to a foliated neighbourhood of the leaf L_o, corresponding to the fixed point 0, of the suspension (Σ_f, F_f). □

3.2.6. - *Remark.* (Notation as in 3.2.5).- It is obvious how to modify the previous proposition when L is a boundary leaf. In this case $\mathbb{R}$ has to be replaced by $\mathbb{R}^+$. Moreover, given an orientation of a neighbourhood of L a homeomorphism between $g(F,L)$ and $g(F_f,L_o)$ may be found to be orientation preserving and arbitrary (orientation preserving) on the boundary.

3.2.7. - *Leaf holonomy*. By $G^r(\mathbb{R},0)$, resp. $G^r(\mathbb{R}^+,0)$ we denote the group of germs of C^r diffeomorphisms which are defined in a neighbourhood of the origin 0 in $\mathbb{R}$, resp. $\mathbb{R}^+$, and which keep 0 fixed. The subgroup $G_+^r(\mathbb{R},0) \subset G^r(\mathbb{R},0)$ consists of all those germs which are represented by an orientation preserving diffeomorphism. Let π be the natural projection from $\mathrm{Diff}^r(\mathbb{R},0)$, resp. $\mathrm{Diff}^r(\mathbb{R}^+,0)$, onto $G^r(\mathbb{R},0)$, resp. $G^r(\mathbb{R}^+,0)$.

Suppose L is a circle leaf of the suspension (Σ_f, F_f) obtained by the representation

$$\chi : \pi_1 S^1 \rightarrow \mathrm{Diff}^r(M), \quad M = \mathbb{R} \text{ or } \mathbb{R}^+,$$

where $\mathrm{im}\,\chi$ is generated by the C^r diffeomorphism f with $f(0) = 0$ and such that L corresponds to 0. We choose an orientation of L and consider the diagram

$$\begin{array}{ccc} \pi_1 L \simeq \pi_1 S^1 & \xrightarrow{\ \chi\ } & \mathrm{Diff}^r(M,0) \\ & \overset{\text{hol}}{\searrow} & \downarrow \pi \\ & & G^r(M,0) \end{array}$$

The homomorphism $\text{hol} = \pi \circ \chi$ is called the <u>holonomy</u> (<u>representation</u>) of L, its image $\text{hol}(L)$ is the <u>holonomy group</u> of $L \in \mathcal{F}_f$.

The following remarks are more or less obvious.

3.2.8. - <u>Remarks</u>. i) If one reverses the orientation of the leaf L the holonomy representation has to be replaced by its inverse.

ii) The holonomy representation depends only on the germ $g(\mathcal{F}_f, L)$, more precisely, on the germ of f in 0.

iii) If two germs of suspensions $g(\mathcal{F}_f, L)$ and $g(\mathcal{F}_{f'}, L')$ are homeomorphic then their holonomy representations are conjugate in $G^0(M,0)$. Conversely, if two holonomy representations are conjugate in $G^0(M,0)$ then the corresponding germs are homeomorphic.

Therefore we will consider the holonomy representation as defined up to conjugation.

iv) Using 3.2.5., it is evident how to define the holonomy of a circle leaf in an arbitrary foliated surface.

v) If the foliation $(\Sigma, \mathcal{F})$ is transversely orientable then the holonomy of a circle leaf in the interior of Σ is always in $G^r_+(\mathbb{R},0)$.

vi) For a boundary leaf L the holonomy will be defined in the obvious way as a representation in $G^r(\mathbb{R}^+,0)$.

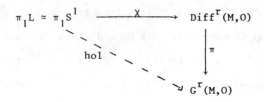

vii) Let L be a two-sided circle leaf of $(\Sigma, \mathcal{F})$. Cutting Σ along L yields a foliation $(\Sigma', \mathcal{F}')$ with two boundary leaves L^+ and L^- coming from L. The holonomy representation

$$\mathrm{hol} : \pi_1(L^+) \to G^r(\mathbb{R}^+,0)$$

of L^+ in $(\Sigma',\mathcal{F}')$ is called the <u>holonomy of</u> L <u>to the right</u> and is denoted by hol^+. In the same way we define hol^-. The corresponding holonomy groups are denoted by $\mathrm{hol}^+(L)$ and $\mathrm{hol}^-(L)$, respectively.

viii) We agree that for simply connected leaves the holonomy is trivial.

3.2.9. - *Exercises*. i) Verify all the statements in 3.2.8.

ii) Construct an example of a foliation with a two-sided leaf L such that $\mathrm{hol}^+(L) \neq 0$ and $\mathrm{hol}^-(L) = 0$.

iii) Let L be a one-sided leaf. Show that hol(L) is isomorphic either to $\mathbb{Z}$ or $\mathbb{Z}_2$. Give an example of a foliation with a one-sided leaf.

3.3. *Reeb components*.

The suspension of a diffeomorphism $f \in \mathrm{Diff}^r_+(I)$ without fixed points in $\mathring{I}$ is a foliation on $S^1 \times I$ which is tangent to the boundary and whose only compact leaves are the boundary curves. The interior leaves of such a foliation approach the boundary leaves in opposite directions, see fig. 14. The two arrowed intervals there have to be identified.

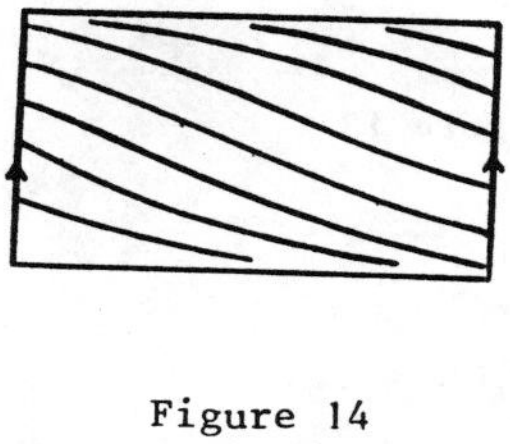

Figure 14

There is another type of foliation on $S^1 \times I$ tangent to the boundary and with all leaves in the interior non compact. This can be described as follows :

In the (x,y)-plane we consider the strip $\Sigma = [-1,1] \times \mathbb{R}$.

For $y \in \mathbb{R}$, let

$$f_y : (-1,1) \to \mathbb{R}$$
$$x \mapsto \frac{x^2}{1-x^2} + y.$$

Then there is a C^∞ foliation $\mathcal{F}_o$ on Σ whose leaves are the boundary lines $x = \pm 1$ and the graphs of f_y, for $y \in \mathbb{R}$. The leaves in $\mathring{\Sigma}$ approach the boundary leaves in the same direction, fig. 15 a). Note that in fig. 6 we had a similar situation.

The foliation $\mathcal{F}_o$ is preserved by the translation $\mathbb{R}_1 : \Sigma \to \Sigma$, $R_1(x,y) = (x,y+1)$. We hence obtain a C^∞ foliation $\mathcal{R}$ on $S^1 \times I$, see fig. 15 b).

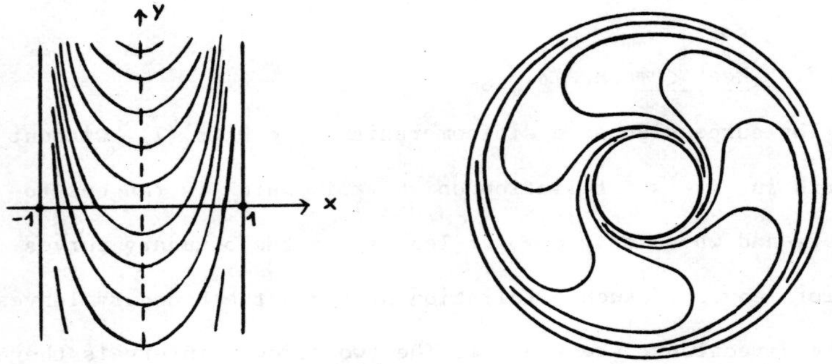

a) b)

Figure 15

The homeomorphism

$$\varphi : \Sigma \to \Sigma$$
$$(x,y) \mapsto (-x,y+1)$$

also preserves the foliation F_o. Therefore we get a non-orientable analogon R_n of R on the Möbius band.

3.3.1. - *Definition.* - A (2-dimensional) Reeb component is any foliation which is homeomorphic to R.

A (2-dimensional) non-orientable Reeb component is any foliation which is homeomorphic to R_n.

A somewhat more intrinsic definition of Reeb components will be given in chapter II. There also higher dimensional Reeb components will be introduced. It is the correspondence between R and the 3-dimensional Reeb component - found by Reeb [Re] - which motivated us to call R a Reeb component although this foliation had been already studied in Kneser [Kn], a paper about thirty years older than [Re].

3.4. *Turbulization.*

3.4.1. - *Definition.* - A closed transversal of a foliated surface (Σ,F) is a simple closed curve θ in Σ without contact points.

Closed transversals will turn out to be a very useful tool for the study of foliations, also in higher dimensions. As we shall see next, they may also be used to construct new foliations out of given ones.

3.4.2. - *Turbulization.* Let θ be a closed transversal of (Σ,F). In a suitable closed neighbourhood U of θ in Σ the foliation $F|U$ is homeomorphic to the trivial foliation by intervals on the annulus or on the Möbius band depending on whether θ is two-sided in Σ or not. This can be seen by taking a covering of θ by finitely many distinguished squares.

Now consider a foliation (Σ_o, F_o) as indicated in fig. 16 a). The two horizontal edges have to be identified either by the identity or by -id, depending on whether Σ_o shall be the annulus or the Möbius band.

We then remove $\overset{\circ}{U}$ from Σ and glue in the foliation (Σ_o, F_o) instead of $F|U$ in such a way that the vertical dotted line in fig. 16 a) corresponds to θ. The surface we obtain by this process is homeomorphic to Σ (see fig. 16 b).

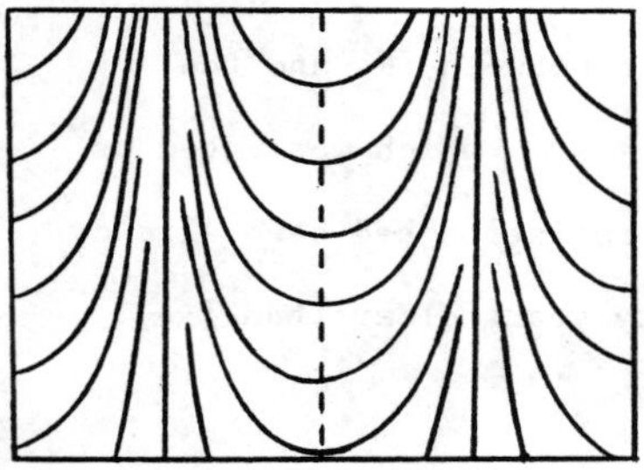

a)

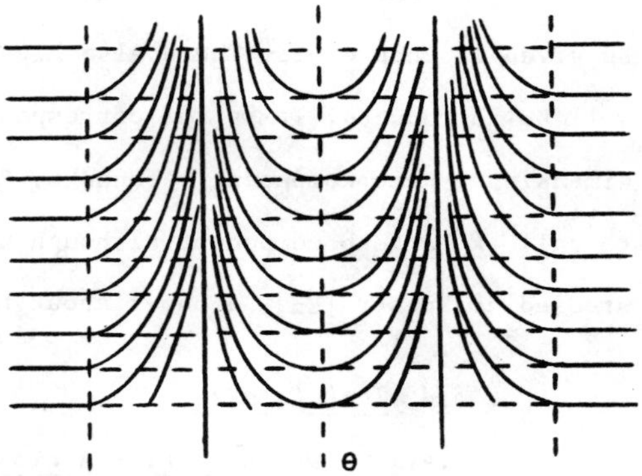

b)

Figure 16

As result of this modification of (Σ, F) we get a foliation (Σ, F') where, roughly speaking, the trivial foliation $F|U$ (dotted in fig. 16 b)) has been replaced by a Reeb component whereas outside U the foliations F and F' coincide.

We say that F' asires from F by turbulization along the closed transversal θ or that F is turbulized, or spiraled, along θ.

The concept of turbulizing a foliation was introduced by Reeb [Re].

For more information about gluing of foliations cf. the next section and § 4. See also exercice 4.2.13.

3.5. *Gluing foliations together.*

Under certain simple conditions foliated surfaces with boundary (Σ, F) and (Σ', F') may be glued together along boundary components to yield a new foliated surface.

Suppose that there are components $\Sigma_o \subset \partial\Sigma$ and $\Sigma'_o \subset \partial\Sigma'$ such that F and F' are either both tangent or both transverse to Σ_o resp. Σ'_o. We may use any homeomorphism $\varphi : \Sigma_o \to \Sigma'_o$ to glue (Σ, F) and (Σ', F') together. This clearly yields a foliation F_1 on the surface $\Sigma_1 = \Sigma \underset{\varphi}{\cup} \Sigma'$.

Note that in general F_1 is not C^r, even if F, F' and φ all are of class C^r, $r \geq 1$, (see the next exercise). But this does not matter since we shall be interested in a classification of C^o foliations up to homeomorphism, as introduced in 2.2.2.

We shall use this gluing process in section 4.2. when gluing of Reeb components with suspensions on $S^1 \times I$ will be used to classify foliations on the annulus and on the Möbius band. Evidently, one has to show in how far the result of gluing depends on the gluing homeomorphism. In fact, as will be shown in 4.2.11, in all cases of interest to us, the gluing process depends only of the isotopy class of the gluing homeomorphism.

If Σ has two boundary components Σ_o and Σ'_o and if F is tangent (resp. transverse) to Σ_o and Σ'_o then we can perform the same gluing process as described above to these two components to obtain a foliation on a closed surface which must be the torus or the Klein bottle.

Exercise. Let $J^s(\mathbb{R},0)$ (resp. $J^s(\mathbb{R}^+,0)$) be the set of s-jets of elements of $G^s(\mathbb{R},0)$ (resp. $G^s(\mathbb{R}^+,0)$), see Hirsch [Hi]. For $s \leq r$, there is a natural projection $J^s : G^r(\mathbb{R},0) \to J^s(\mathbb{R},0)$. We denote by $hol : \pi_1 L \to G^r(\mathbb{R},0)$ the holonomy of the circle leaf L of the C^r foliation F. Then

$$J^s \circ hol : \pi_1 L \to J^s(\mathbb{R},0)$$

is the <u>infinitesimal holonomy of L of order</u> s, (defined up to conjugation). For boundary leaves and for the holonomy of L to the left or to the right the definition of infinitesimal holonomy is analogous.

Now let (Σ,F), (Σ',F'), $\varphi : \Sigma_o \to \Sigma'_o$ and (Σ_1,F_1) be as above and let $L \in F_1$ be the leaf coming from Σ_o and Σ'_o. Prove that if F,F' and φ are of class C^r then F_1 is of class C^r if and only if $J^r hol^+(L) = J^r hol^-(L)$.

4. *Classification of foliations on compact surfaces.*

4.1. *Topological dynamics.*

4.1.1. We already know that the leaves of a foliated surface are the familiar 1-dimensional manifolds. A 1-dimensional manifold may however appear in different ways as a leaf of a foliation. To make this clear we take a leaf L of a foliation (Σ,F) and pick two points x_o,x_1 on L. Let c be a curve in L joining x_o and x_1 and let U be a distinguished square containing c, see 3.2.4. Then the foliation in U defines a local homeomorphism

$$\varphi : (c_o,x_o) \to (c_1,x_1)$$

between transversals c_o and c_1 of F in x_o and x_1, respectively, and mapping $c_o \cap L$ in $c_1 \cap L$. This permits us to describe the local transverse behaviour of a leaf L by means of a transversal through any point of L.

4.1.2. Leaf types. Using transversals we may partition the leaves of F into three types :

i) There exists a transversal c_o through $x_o \in L$ such that $c_o \cap L = \{x_o\}$. Then, by our introducing remarks, the topologies of L induced by the topology of Σ and by the leaf topology coincide. We say that L is proper. Note that a closed leaf is proper.

ii) If L is not proper then, for any $x_o \in \bar{L}$, there is a transversal c_o passing through x_o such that $c_o \cap \bar{L}$ is homeomorphic to a perfect subset of $\mathbb{R}$, (that is to say a closed set without isolated points). There are two kinds of such sets.

ii_1) $c_o \cap \bar{L}$ is a closed interval. In this case we say that L is locally dense.

ii_2) $c_o \cap \bar{L}$ is a Cantor set (see 3.1.4, ii) c)). The leaf L is then exceptional.

4.1.3. Definition. - Let (Σ, F) be a foliated surface. A subset $A \subset \Sigma$ is saturated (or invariant) if for every $a \in A$ the leaf through A is contained in A.

We denote by $\mathring{A}$ the interior of A in Σ.

4.1.4. Lemma. - If $A \subset (\Sigma, F)$ is saturated then $\bar{A}, \mathring{A}$ and $\bar{A} - \mathring{A}$ are also saturated.

Proof : The lemma is an immediate consequence of 4.1.1. □

4.1.5. *Definition.* - By a minimal set of the foliation (Σ,F) we mean a non-empty saturated closed subset M of (Σ,F) which is minimal with respect to inclusion, (i.e. if $M' \subset M$ is non-empty saturated and closed then $M' = M$).

Minimal sets are characterized by the following proposition.

4.1.6. *Proposition.* - A non-empty saturated set $M \subset (\Sigma,F)$ is minimal if and only if for every leaf $L \subset M$ we have $\bar{L} = M$.

Proof : If M is minimal then $\bar{L} = M$, by 4.1.4. To prove the converse let $M' \subset M$ be non-empty closed and saturated. For $L \subset M'$ we have $M = \bar{L} \subset M'$, by our assumption. Therefore $M' = M$ and thus M is minimal. □

It is now clear that all leaves contained in a minimal set are of the same type. Moreover, a proper leaf is a minimal set if and only if it is closed (i.e. compact if Σ is compact). We have hence three types of minimal sets.

4.1.7. *Types of minimal sets.* A minimal set $M \subset (\Sigma,F)$ satisfies one of the following three possibilites :

(1) M is a proper (compact) leaf ;

(2) M is all of Σ ;

(3) M is a nowhere dense set which is the union of exceptional leaves. In this case M is called an exceptional minimal set.

4.1.8. *Proposition.* - Let (Σ,F) be a compact foliated surface. Then $\bar{L}$ contains a minimal set, for every $L \in F$. In particular, (Σ,F) contains a minimal set.

Proof : Let L be a leaf of F. Since Σ is compact $\bar{L}$ is also compact. Therefore every decreasing family $\{A_j\}$ of non-empty closed saturated sets in $\bar{L}$ has a non-empty intersection. The result follows from Zorn's lemma. □

To conclude this section we give a brief study of the union of all minimal sets of a compact foliated surface.

First recall the obvious fact that the compact leaves of a suspension (Σ_f, F_f) of a homeomorphism f of $M = I$ or S^1 are in one-one correspondence with the finite orbits of f . Furthermore, all finite orbits of f are of the same order (cf. 5.1.6) and therefore their union is closed. This shows that the union of compact leaves of F is closed.

4.1.9. Lemma. - If (Σ_f, F_f) is as above and if F_f contains an exceptional minimal set $\mathcal{M}$ then

(1) $\mathcal{M}$ is the unique minimal set of F_f and

(2) f is an orientation preserving homeomorphism of S^1 (and therefore $\Sigma_f = T^2$).

Proof : By construction (see 3.1.2.) Σ_f is the total space of an M-bundle over S^1, where $M = I$ or S^1. If we identify M with the fiber over $1 \in S^1$ then for every $x \in M$ the intersection $M \cap L_x$ coincides with the orbit of x under f. Moreover, by 4.1.2., the intersection $M \cap \mathcal{M}$ is a Cantor set invariant under f and $\mathrm{cl}(M \cap L_x) = M \cap \mathcal{M}$. When $M = I$ all orbits of f would have isolated points which is impossible. Therefore M must be S^1.

Let J be the closure of a component of $M - \mathcal{M}$. We may think of J as a closed interval $[x_0, x_1]$ with $x_1 \in \mathcal{M}$.

As L_{x_o} is exceptional there exists a sequence $\{\psi(n)\}$ such that $\lim_{n\to\infty} f^{\psi(n)}(x_o) = x_o$. This clearly implies that

$$\lim_{n\to\infty} f^{\psi(n)}(x) = x_o \quad \text{for any} \quad x \in J.$$

This argument shows that $\bar{L} \supset M$ for every leaf L of F_f. By 4.1.6., M is the only minimal set of F_f. We conclude that f cannot have a periodic orbit and thereforemust be orientation preserving. But this means $\Sigma_f = T^2$. □

The next theorem gives a more detailed information about leaf types and minimal sets. It will be a consequence of the next two sections, cf. 4.2.16 and 4.3.6.

4.1.10. *Theorem*. - Let (Σ, F) be a compact foliated surface.

i) If F admits a compact leaf then all leaves are proper and the union of compact leaves is compact.

ii) If F contains a locally dense leaf then all leaves are dense in Σ, F is a suspension and $\Sigma = T^2$.

iii) If F admits an exceptional minimal set this minimal set is unique, F is a suspension and $\Sigma = T^2$.

4.1.11. *Exercise*. - Decide whether or not proposition 4.1.8. is true also for open foliated surfaces.

Remark. It is worth noting that leaf types and the type of a minimal set are both invariant under homeomorphism.

4.2. *Foliations on the annulus and on the Möbius band.*

In this section we shall give a classification of foliations on the annulus $S^1 \times I$ and on the Möbius band. This classification will be up to homeomorphism, as defined in 2.2.2. and we shall not be interested in the differentiability class of foliations.

4.2.1. *Remark.* Let A be either the annulus or the Möbius band and let F be a foliation on A which is transverse to the boundary component A_o of A. Then from F we obtain a foliation (A,F') tangent to A_o if we let F "spiral" along A_o. By this we mean the following.

In a small product neighbourhood U of A_o in A the foliation $F|U$ is homeomorphic to the product foliation on $S^1 \times [0,1]$ by intervals. According to 3.4. we may replace $F|U$ by "a half Reeb component" , for example by $R|S^1 \times [0,1]$, where R is the Reeb component defined in 3.3.

Clearly there are two essentially different ways we can do this "spiraling" of F, depending on whether the gluing homeomorphism preserves orientations or not. Fig. 17 indicates the two possibilities that one can get as F_1 when the product foliation F on $S^1 \times I$ is spiraled along the boundary circle $S^1 \times \{1\}$.

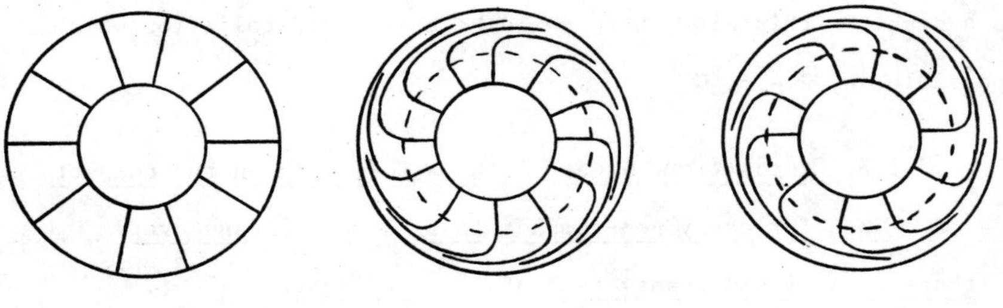

Figure 17

Conversely, if F is a foliation on A which is tangent to the boundary circle A_o and whose holonomy group in A_o is generated by a germ of $G^r(\mathbb{R}^+,0)$ without fixed points different from 0 then near A_o there is a closed transversal c of F. If we cut (A,F) along c we see that F may be considered as obtained by spiraling a foliation F' on A which is transverse to A_o.

This shows that there is a close relationship between foliations tangent to ∂A and those transverse to ∂A. This is the reason why we only need to study foliations tangent to ∂A. It will turn out that the suspensions and the Reeb components may serve as building blocks for all other foliations.

Our next aim is to show that on $S^1 \times I$ there are exactly two isomorphism classes and on the Möbius band there is exactly one isomorphism class of foliations which are tangent to the boundary and without circle leaves in the interior, see 4.2.12. To this end we shall first prove the two following propositions each of which is also of some interest in itself.

4.2.2. *Proposition*. - A closed transversal or a circle leaf cannot be homotopic to zero.

Proof. On surfaces every simple closed null homotopic curve bounds a disk. On this disk there would be an induced foliation which is impossible, by 2.4.6. □

4.2.3. *Proposition*. - Let F be a foliation on the compact surface Σ. Then for every non-compact leaf L of F and every $x \in L$ there is a closed transversal of F passing through x.

Proof : First suppose that F is transversely orientable. Since Σ is compact and L is non-compact there exists a distinguished open square U of F such that two different plaques P_1, P_2 of U lie in L. We choose points $x_1 \in P_1$ and $x_2 \in P_2$ and join them in U by a transversal c.

The arc b on L with x_1 and x_2 as endpoints cannot intersect U infinitely often. We may therefore assume that c does not meet $\overset{\circ}{b}$ at all, fig. 18 a). Now the arcs c and b together form a closed curve σ in Σ. Since F is transversely orientable σ can be approximated by a closed transversal θ. This can be seen by covering the arc b with a distinguished open square U', see 3.2.4. and fig. 18 b).

When F is not transversely orientable we take the two-fold transverse orientation covering $\pi : \Sigma^* \to \Sigma$, see 2.3.4., with γ the non-trivial covering translation.

For $L \in F$ let $L^* \in \pi^*F$ such that $\pi(L^*) = L$. We construct a closed transversal θ_o through L^*, as in the first part of the proof. Then θ_o and $\gamma(\theta_o)$ intersect in at most a finite number of points. We shall modify θ_o and $\gamma(\theta_o)$ equivariantly so that they become disjoint.

Let y be an intersection point of θ_o and $\gamma(\theta_o)$. In a neighbourhood of y we modify θ_o and $\gamma(\theta_o)$ equivariantly with respect to γ so that there are two points of intersection less between θ_o and $\gamma(\theta_o)$. This is indicated in fig. 18 c). After finitely many steps θ_o and $\gamma(\theta_o)$ have been replaced by a collection $\sigma_1, \gamma(\sigma_1), \ldots, \sigma_m, \gamma(\sigma_m)$ of closed transversals. We may suppose without loss of generality that σ_1 intersects L^*.

Now either $\sigma_1 = \gamma(\sigma_1)$ or $\sigma_1 \cap \gamma(\sigma_1) = \emptyset$. In both cases $\theta = \pi(\sigma_1)$ is a closed transversal through $L \in F$.

Using again 3.2.4. it is easily seen (fig. 18 d)) that θ may be modified to pass through the point x.

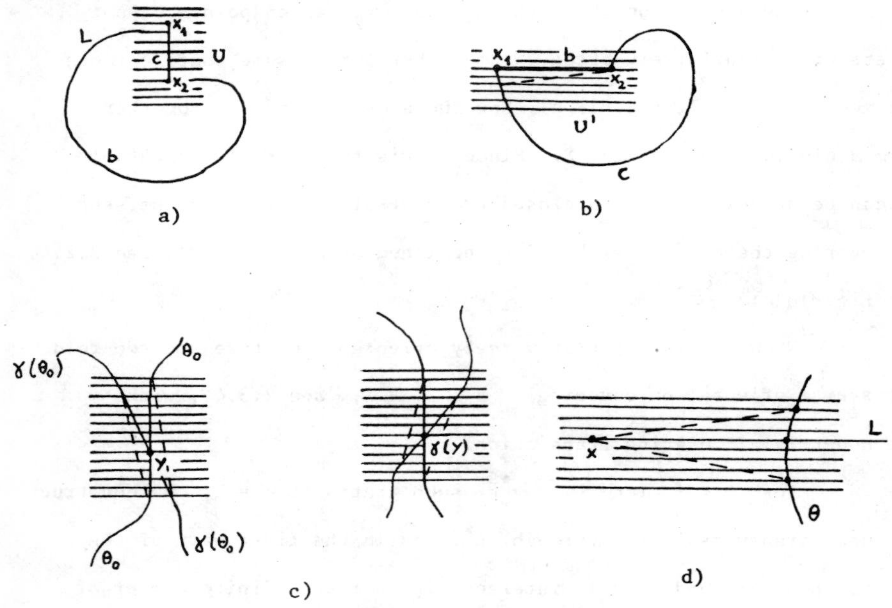

Figure 18

4.2.4. Lemma. - Let A be either the annulus or the Möbius band and let F be a foliation on A. Then every minimal set of F is a compact leaf.

Proof : We may suppose that F is transversely orientable and that $A = S^1 \times I$. Otherwise we could pass to the 4-fold covering (A^*, F^*) which makes A orientable and F transversely orientable. The lemma is then true for (A, F) if and only if it is true for (A^*, F^*).

Let L be a non-compact leaf of F and θ a closed transversal through L. Since θ cannot be homotopic to zero it must be isotopic to the boundary curves of $A = S^1 \times I$. As F is transversely orientable it is also orientable, by 2.3.10 and 2.3.11. Thus the leaf L cannot intersect θ in more than one point and the lemma follows from 4.1.2. and 4.1.7. □

4.2.5. *Lemma.* - Let A be either the annulus or the Möbius band and let F be a foliation on A, without circle leaves. Then F is isotopic to the fibration $I \hookrightarrow A \to S^1$. In particular, F is transversely orientable.

Proof. By 4.1.8 and 4.2.4., every leaf contains a compact leaf in its closure. Since there are no circle leaves this leaf must be an interval. As the set of interval leaves is closed the lemma follows from 3.2.3. □

4.2.6. *Definition.* - Let J be either the open or the closed unit interval. We say that an orientation preserving homeomorphism $f : J \to J$ is above (below) the diagonal if $f(x) > x$ (resp. $f(x) < x$) for all $x \in \mathring{J}$.

4.2.7. *Lemma.* - Let $J = I$ or $\mathring{I}$ and let $f, g : J \to J$ be two homeomorphisms above (resp. below) the diagonal. Then f and g are conjugate in $\mathrm{Homeo}_+(J)$, i.e. there is $h \in \mathrm{Homeo}_+(J)$ such that $h \circ f = g \circ h$.

Proof : First observe that for all $t \in J$ necessarily

$$\lim_{n\to\infty} f^n(t) = \lim_{n\to\infty} g^n(t) = 1$$

and $$\lim_{n\to-\infty} f^n(t) = \lim_{n\to-\infty} g^n(t) = 0.$$

Now we fix s_o and t_o in $\mathring{J}$ an choose an arbitrary orientation preserving homeomorphism

$$h_o : [s_o, f(s_o)] \to [t_o, g(t_o)].$$

Finally, we define $h : J \to J$ by $h(0) = 0$ and $h(1) = 1$, if $J = I$, and

$$h(s) = g^n \circ h_o \circ f^{-n}(s), \quad \text{for} \quad s \in [f^n(s_o), f^{n+1}(s_o)].$$

If f and g are below the diagonal then there is h such that $h \circ f^{-1} = g^{-1} \circ h$ and thus $h \circ f = g \circ h$. □

It is easily seen that a Reeb component and a suspension on $S^1 \times I$ cannot be homeomorphic. For a suspension there always exists a transversal joining the two boundary circles whereas for a Reeb component such a transversal does not exist.

The next three results state in particular that, up to isomorphism, there are exactly two foliations on $S^1 \times I$ tangent to the boundary and without circle leaves in the interior.

4.2.8. *Lemma*. - Let $\mathcal{F}_1$ and $\mathcal{F}_2$ be suspensions on $S^1 \times I$ obtained by the homeomorphisms f_1 and f_2. If both $\mathcal{F}_1$ and $\mathcal{F}_2$ have no circle leaves in the interior then they are conjugate.

Proof : As $\mathcal{F}_i$ has no interior circle leaves it follows that f_i is either attracting or expanding. Suppose first that f_1 and f_2 are expanding. Then by 4.2.7., there is a homeomorphism h of I such that $h \circ f_1 = f_2 \circ h$. Let $\pi_i : \mathbb{R} \times I \to S^1 \times I$ be the covering projection, as used in the definition of F_i, $i = 1,2$, see 3.1.2. We define $\tilde{H} : \mathbb{R} \times I \to \mathbb{R} \times I$ by $\tilde{H}(x,t) = (x,h(t))$. Then there is a unique homeomorphism $H : S^1 \times I \to S^1 \times I$ such that the following diagram is commutative

$$\begin{array}{ccc} \mathbb{R} \times I & \xrightarrow{\tilde{H}} & \mathbb{R} \times I \\ \downarrow{\scriptstyle \pi_1} & & \downarrow{\scriptstyle \pi_2} \\ S^1 \times I & \xrightarrow{H} & S^1 \times I \end{array}$$

By construction, H takes F_1 onto F_2. Moreover, as $\tilde{H}$ is the identity on the boundary it follows that H is also the identity on the boundary.

When f_1 and f_2 are below the diagonal, the construction of H is similar, also with the identity on the boundary. When, say f_1 is below and f_2 is above the diagonal, then we have to replace f_1 by f_1^{-1} and the above argument yields a homeomorphism between the suspension F_1' of f_1^{-1} and F_2, also the identity on the boundary. The proof is completed by the observation that F_1 and F_1' are isomorphic. □

The proof of 4.2.8. evidently shows more than we claimed, namely :

4.2.9. *Corollary*. - Let F_1 and F_2 be two suspensions on $S^1 \times I$ obtained by homeomorphisms of I which are above (resp. below) the diagonal. Then F_1 and F_2 are conjugate by a homeomorphism which is the identity on the boundary.

4.2.10. *Lemma*. - Let (Σ, F) be a foliation on $S^1 \times I$ or on $S^1 \times_{\mathbb{Z}_2} I$ tangent to the boundary. Then F is conjugate to a suspension if and only if there exists a transversal c such that $c \cap \partial\Sigma = \partial c$.

Proof : Let c be a transversal such that $c \cap \partial\Sigma = \partial c = \{x_o, x_1\}$. We identify points of $\partial\Sigma$ by means of an orientation preserving homeomorphism which takes x_o to x_1 and which should be of order two and without fixed points if $\partial\Sigma$ is connected. The result is a foliation (Σ', F') where $\Sigma' = T^2$ or K^2 depending on whether Σ is orientable or not. On F' there is a closed transversal θ coming from c. If we cut Σ' along θ we get an annulus A together with a foliation F_o transverse to ∂A. Each circle leaf of F must intersect the curve c, by 4.2.2. Thus we may apply 4.2.5 to see that (A, F_o) is a fibration. It follows that $(S^1 \times I, F)$ is homeomorphic to a suspension.

The other direction is trivial and so the lemma is proved. □

4.2.11. Lemma. - Let $(S^1 \times I, R_o)$ be the right half of the Reeb component R, as introduced in 3.3.1., i.e. $R_o = R|(S^1 \times [0,1])$. Denote by A_i the boundary curve $S^1 \times \{i\}$, $i = 0, 1$. Given an orientation preserving homeomorphism $h : A_i \to A_i$, $i = 0$ or 1, there exists a homeomorphism $H : (S^1 \times I, R_o) \to (S^1 \times I, R_o)$ with $H|A_i = h$ and $H|A_{1-i} = id$.

Proof : We may take as model of R_o a foliation which is transverse to the fibration F_o by intervals $\{x\} \times I$. Let F_1 be another fibration of $S^1 \times I$ by intervals transverse to R_o but such that $(x, 1-i)$ is joined to $(h(x), i)$. We are going to construct a homeomorphism H such that $H|A_i = h$, $H|A_{1-i} = id$ and H takes (R_o, F_o) onto (R_o, F_1).

For $p = (x,t) \in S^1 \times (0,1)$ the leaf $L_p \in R_o$ meets A_o in the point y. The leaf $L_{H(y)} \in R_o$ intersects the leaf of F_1 through $(h^i(x), 1)$ the first time in $p' \in S^1 \times (0,1)$. We set $H(p) = p'$.

Then H has the required properties. □

4.2.12. *Proposition*. - Let (Σ, F) be a foliation tangent to the boundary and without circle leaves in the interior of Σ.

i) If $\Sigma = S^1 \times I$ then F is either a Reeb component or it is homeomorphic to a suspension.

ii) If $\Sigma = S^1 \times_{\mathbb{Z}_2} I$ then F is a non-orientable Reeb component.

Proof : We first consider $S^1 \times I$. As there are no circle leaves in the interior we may find a closed transversal θ near the boundary curve $S = S^1 \times \{0\}$. If $A \subset \Sigma$ denotes the annulus bounded by S and θ the foliation $F|A$ is homeomorphic to the half Reeb component R_o, according to 3.2.6 and 4.2.8. The same argument, together with 4.2.5., shows that $F|cl(\Sigma-A)$ is also homeomorphic to R_o. If homeomorphisms between $F|A$ resp. $F|cl(\Sigma-A)$ and R_o can be found which preserve orientation then it follows by 4.2.11 that F is a Reeb component. Otherwise we can easily find a transversal of F connecting the boundary curves and an application of 4.2.10 shows that F is homeomorphic to a suspension.

If $\Sigma = S^1 \times_{\mathbb{Z}_2} I$ we may find an annulus A as above and $F|A$ is again homeomorphic to R_o whereas $F|cl(\Sigma-A)$ is a fibration by intervals. Now consider the non-orientable Reeb component R_n. In a closed annular neighbourhood of the core R_n is a fibration by intervals and we may take a homeomorphism $h : F|cl(\Sigma-A) \to R_n|A'$ that should preserve the orientations of the boundary if and only if the homeomorphism between $F|A$ and R_o can be chosen to be orientation preserving. Applying 4.2.11 once more, we see that h can be extended to a homeomorphism between F and R_n, as required. □

4.2.13. Exercises. i) Let F_1 and F_2 be foliations on the Möbius band which are obtained from Reeb components on $S^1 \times I$ by identification on one boundary component by means of a fixed point free involution. Show that F_1 and F_2 are homeomorphic.

ii) Show that turbulization depends only on the isotopy class (i.e. here on the orientation behaviour) of the gluing homeomorphisms.

iii) Let θ_o and θ_1 be closed transversals in the interior of a foliated surface. If θ_o and θ_1 are isotopic through closed transversals prove that turbulization along θ_o and θ_1 by means of orientation preserving (resp. reversing) homeomorphisms yields homeomorphic results.

We now come to the main result of this section, that is the classification of foliations on the annulus and on the Möbius band which are tangent to the boundary. Its proof is based on 4.2.11., 4.2.12. and the next lemma.

4.2.14. Lemma. Let Σ be either the annulus or the Möbius band and let C be the union of all closed leaves of a foliation F on Σ tangent to the boundary. Then

i) C is closed,

ii) F contains at most finitely many Reeb components.

Proof : It suffices to prove the lemma for $\Sigma = S^1 \times I$ and F transversely orientable.

Let $x \in \bar{C}$ then the leaf L_x contains a minimal set M in its closure and $M = L \in F$, by 4.2.4. The germ $g(F,L)$ is homeomorphic to a germ of a suspension; see 3.2.5. Therefore if L_x is not compact it spirals towards L on one side, let us say on the right. Then clearly

L is isolated on the right in C. But this is impossible.

Property ii) holds since otherwise we could find a transversal of F intersecting all leaves of a Reeb component contradicting 4.2.10. □

Remark. (Notation as in 4.2.14).- Note that there are foliations on Σ such that the intersection of C with a suitable transversal is a Cantor set.

4.2.15. Theorem. - Up to homeomorphism we have :

i) Every foliation on $S^1 \times I$ tangent to the boundary is obtained by gluing together a finite number of Reeb components and a finite number of suspensions.

ii) Every foliation on $S^1 \times_{\mathbb{Z}_2} I$ tangent to the boundary is one of the following a) - c), possibly glued together with a foliation on $S^1 \times I$.

a) The non-orientable Reeb component.

b) The (orientable) Reeb component identified on one boundary circle by means of a fixed point free involution.

c) A suspension of an orientation reversing self-homeomorphism of the interval.

Furthermore, the result of gluing depends only on the isotopy class of the gluing homeomorphisms.

Also, every foliation on $S^1 \times I$ is transversely orientable and on $S^1 \times_{\mathbb{Z}_2} I$ a foliation is transversely orientable if and only if there is no one-sided circle leaf.

Proof : We first consider the annulus. By 4.2.14., ii), every foliation $(S^1 \times I, F)$ contains at most finitely many Reeb components $R_1, \ldots, R_m$. Let $(A_o, F|A_o)$ be the closure of a component of $F - \cup R_i$. We claim that $F|A_o$ is homeomorphic to a suspension.

For this it suffices to show, by 4.2.11., that there exists a transversal of $F|A_o$ joining the two boundary curves. We first cover A_o with finitely many closed distinguished squares (U_j,φ_j), $j = 1,\dots,m$. Then $c_j = \varphi_j^{-1}(\{0\} \times [-1,1])$ is a transversal and each leaf of $F|A_o$ intersects some c_j.

Let C be the union of the compact leaves in A_o. If one of the endpoints of c_j lies in a component of $A_o - C$ then, using 4.2.10, we may extend c_j to a transversal so that this endpoint comes to lie on a compact leaf. Thus we may suppose that all endpoints of $c_1,\dots,c_m$ lie on compact leaves.

Denote by (A_j,F_j) the foliation formed by the leaves intersecting c_j, $j = 1,\dots,m$. There is one of the c_j, say c_2, such that $c_2 \cap A_1 \neq \emptyset$. We may apply 3.2.4. to find a transversal c of $F|(A_1 \cup A_2)$ which meets every leaf of $F|(A_1 \cup A_2)$. Now this process is repeated with $c,c_3,\dots,c_m$ instead of $c_1,\dots,c_m$. After finitely many steps we get a single transversal intersecting all leaves of $F|A_o$. Applying 4.2.10 once more, we see that $F|A_o$ is a suspension.

Since there are only finitely many R_i there cannot be infinitely many components in $F - \cup R_i$. Thus i) is proved.

Now let $(S^1 \times_{\mathbb{Z}_2} I,F)$ be given and let C be the union of the circle leaves in F. If F does not contain a one-sided leaf then the closure of exactly one component of $F - C$ is a non-orientable Reeb component, by 4.2.12. Hence in this case F is obtained by gluing together a foliation on $S^1 \times I$ and a non-orientable Reeb component.

If F contains a one-sided leaf L - there can be at most one such - then we cut F along L. This yields a foliation F_o on $S^1 \times I$ which may be decomposed as in i). Therefore the boundary leaf

L_o of F_o which originates from L either lies in a Reeb component or in a suspension. If it lies in a Reeb component then L is contained in a component of type b). Otherwise L lies in a suspension of an orientation reversing selfhomeomorphism of I, i.e. c) holds. This proves ii).

It follows from 4.2.11 (together with the fact that two self-homeomorphisms of S^1 are isotopic if and only if they are orientation preserving) that all identifications between boundary leaves depend only on the orientation behaviour of the gluing homeomorphisms. (Hence we can always take id_{S^1} or $-\overline{id_{\mathbb{R}}}$).

The two Reeb components and all suspensions on $S^1 \times I$ are transversely orientable. Gluing together two such foliations along a boundary leaf gives again a transversely orientable foliation. We conclude that all foliations on $S^1 \times I$ are transversely orientable.

A foliation F on the Möbius band cannot be transversely orientable if it contains a one-sided leaf. On the other hand if there is no one-sided leaf in F then F is the union of a non-orientable Reeb component and a foliation on $S^1 \times I$. Thus F is transversely orientable. □

4.2.16. *Remarks*. i) Note that in 4.2.15. the decomposition of a foliation is unique if the suspension components are taken to be maximal, that is any two of them are disjoint.

ii) Together with 4.2.14. theorem 4.2.15. proves theorem 4.1.10. for Σ the annulus or the Möbius band.

Exercise. Using the notion of holonomy of circle leaves describe all topologically distinct foliations on the Möbius band with three compact leaves.

4.3. *Foliations on the torus and on the Klein bottle.*

We shall distinguish between foliations without compact leaves and foliations with at least one compact leaf. The latter will be classified using the results on the annulus and the Möbius band. We begin with the investigation of suspension foliations. By a suspension we mean in future a suspension, up to homeomorphism.

4.3.1. Lemma. - Let (Σ,F) be a foliated torus or Klein bottle. The two following conditions are equivalent :

i) F is the suspension of a homeomorphism of S^1.

ii) There exists a closed transversal of F intersecting every leaf.

Proof : Obviously, i) implies ii). To prove the converse we first suppose that F does not contain a compact leaf. Let (Σ_o,F_o) be the foliated surface obtained by cutting Σ along the closed transversal θ. Then Σ_o is either an annulus, a Möbius band or it consists of two Möbius bands. As θ intersects every leaf of F the foliation F_o is determined by 4.2.5. Thus, if Σ_o consists of two Möbius bands we may take as new closed transversal θ_1 the core of one of them. Cutting Σ along θ_1 yields a foliation (Σ_1,F_1) with Σ_1 a single Möbius band and F_1 is again determined by 4.2.5.

If Σ_o is an annulus we are done. Otherwise we consider (Σ_1,F_1). There is a transversal c in Σ_o such that

(1) $\partial c = c \cap \partial\Sigma_o$,

(2) after reidentification of Σ_o to Σ, c becomes a closed transversal θ_2 of F,

(3) cutting Σ along θ_2 gives an annulus A, cf. fig. 19.

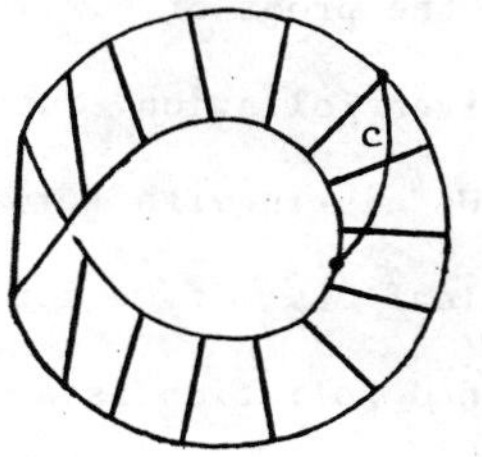

Figure 19

The foliation (A,F_2) obtained from (Σ,F) by cutting along θ_2 is also a foliation by intervals because there are no circle leaves in F. We conclude that also in this case F is a suspension.

Suppose now that F contains a circle leaf L. Cutting Σ along L yields a foliation (Σ_o,F_o) tangent to the boundary and Σ_o is again either an annulus, a Möbius band or the union of two Möbius bands. As there exists a closed transversal intersecting all leaves of F we may apply 4.2.10 to see that each component of (Σ_o,F_o) is a suspension. We conclude that (Σ,F) is a suspension. □

4.3.2. *Proposition*. - Let (Σ,F) be a foliated torus or Klein bottle. Then we have one of the two exclusive situations :

(1) F is a suspension or

(2) F contains a Reeb component (orientable or not).

Proof : We show that F is a suspension if and only if it does not contain a Reeb component. This is trivial in one direction.

As for the other implication we first remark that the same statement holds for Σ the annulus or the Möbius band and F a foliation on Σ tangent to the boundary. Indeed, this can be seen by 4.2.10 in connection with an argument used in the proof of 4.2.15.

Now we come back to the given foliation. Suppose that F does not contain a Reeb component. We distinguish between two cases:

a) F contains a closed leaf L. If we cut Σ along L then each component of the so obtained foliation is a suspension, by the above remark. On each component we choose a transversal intersecting all leaves. These transversals may be used to construct, by means of 3.2.4., a closed transversal of F which intersects every leaf of F.

b) All leaves of F are non-compact. We may then take a closed transversal, according to 4.2.3., and the result follows from 4.2.5. and 4.3.1. □

The last proposition may be applied to prove the

4.3.3. *Theorem*. - Let (Σ,F) be a foliation on the torus or Klein bottle without compact leaves. Then F is the suspension of a homeomorphism f of S^1. Moreover :

(i) f is orientation preserving and $\Sigma = T^2$.

(ii) Either all leaves of F are everywhere dense or there exists a unique exceptional minimal set M with all leaves of $\Sigma - M$ being proper.

Proof : By 4.3.2., F is a suspension of a homeomorphism f of S^1. If f were orientation reversing then it would have a periodic point and thus F would have a compact leaf. This proves (i).

The second assertion is a consequence of 4.1.9. □

From 4.3.3. we deduce Kneser's theorem [Kn] :

4.3.4. Theorem. - Any foliation on the Klein bottle has a compact leaf.

We conclude this paragraph with a description of all foliations on the torus T^2 and on the Klein bottle K^2 having at least one Reeb component. Together with theorems 2.4.6, 4.2.15 and 4.3.3 this provides a classification of all foliations on all compact surfaces.

4.3.5. Theorem. - Let $\mathcal{F}$ be a foliation on $\Sigma = T^2$ or K^2 which contains at least one Reeb component. Then all leaves of $\mathcal{F}$ are proper and we have :

i) If $\Sigma = T^2$ then $\mathcal{F}$ is obtained by identifying the boundary leaves of a foliation on the annulus by an orientation preserving homeomorphism.

ii) If $\Sigma = K^2$ then $\mathcal{F}$ is obtained by one of a) - c) :

a) Gluing together two foliations on Möbius bands by a homeomorphism which may be orientation preserving or not.

b) Identification of the boundary leaves of a foliation on the annulus by an orientation reversing homeomorphism.

c) Identification on the boundary leaf of a foliation on the Möbius band by means of a fixed point free involution.

Furthermore, the result of the gluing process depends only on the isotopy class of the gluing homeomorphism.

Proof : Let L be a leaf of $\mathcal{F}$ in the boundary of a Reeb component. The surface Σ_o obtained by cutting Σ along L is either an annulus, a Möbius band or the union of two Möbius bands. Clearly, if $\Sigma = T^2$ then Σ_o must be an annulus and the gluing homeomorphism must be orientation preserving. Thus i) and ii) are proved.

The supplementary statement follows by 4.2.11. ◻

4.3.6. *Remark.* Theorem 4.3.5., together with 4.3.3. proves theorem 4.1.10 for $\Sigma = T^2$ and $\Sigma = K^2$.

4.3.7. *Exercises.* i) Give a necessary and sufficient condition for foliations on T^2 resp. K^2 to be (transversely) orientable.

ii) Which are the homology classes in $H_1(K^2;\mathbb{Z})$ that can be represented by a compact leaf of a foliation of K^2 ?

iii) Let (Σ,F) be a foliation on T^2 or K^2 with all leaves compact.

a) If $\Sigma = T^2$ show that F is homeomorphic to a product foliation.

b) Classify all such foliations on K^2, up to conjugation.

iv) Prove that, up to conjugation, there are countably many foliations on $S^1 \times I$, $S^1 \times_{\mathbb{Z}_2} I$ and K^2 but uncountably many on T^2.

v) A foliation is analytic when all transition maps are real analytic. Let F be an analytic foliation on $S^1 \times I$ or $S^1 \times_{\mathbb{Z}_2} I$ tangent to the boundary.

a) Show that either F has finitely many compact leaves or all leaves of F are compact.

b) Can a foliation as indicated in fig. 20 be analytic ?

c) What can be deduced from a) and b) about analytic foliations on T^2 or K^2 ?

vi) By lifting foliations on T^2 or K^2 to their universal covering one gets foliations on $\mathbb{R}^2$. Here is a brief description of foliations $(\mathbb{R}^2,F)$:

a) If F is of class C^1, it is defined by a non-singular complete vector field.

b) Every leaf L of F is closed and $\mathbb{R}^2 - L$ has two components.

c) Any transversal of F intersects any leaf of F in at most one point.

d) The quotient space $B = \mathbb{R}^2/F$ arising from identifying the leaves of F to points and endowed with the quotient topology is a one-dimensional simply connected (and in general non-Hausdorff) manifold and the projection $\pi : \mathbb{R}^2 \to B$ is a locally trivial fibration.

Determine the space B for the foliation indicated in figure 21.

Figure 20

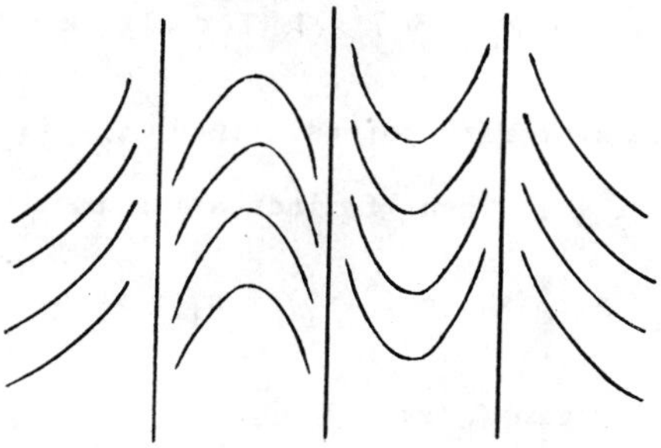

Figure 21

5. *Denjoy theory on the circle.*

Other than on compact surfaces with boundary there are uncountably many topologically distinct foliations on the torus. To see this we use the close relationship between foliations on the torus without Reeb components and diffeomorphisms of the circle.

5.1. *The rotation number.*

With every orientation preserving homeomorphism of S^1 there is associated a real number. This number will turn out to be invariant under conjugation and thus provides a powerful algebraic invariant for suspension foliations on T^2.

5.1.1. *Proposition.* - If $f \in D^o(S^1)$ then there exists a number $\rho(f) \in \mathbb{R}$ with

$$\rho(f) = \lim_{n\to\infty} \frac{f^n(x)}{n} \quad \underline{\text{for all}} \quad x \in \mathbb{R}.$$

Proof : Let $\psi_k = f^k - \text{id}$, for $k \in \mathbb{N}$. Then ψ_k is $\mathbb{Z}$-periodic and therefore has minimum and maximum. Set

$$m_k = \frac{1}{k} \min_x \psi_k(x) \quad \text{and} \quad M_k = \frac{1}{k} \max_x \psi_k(x).$$

We claim that

$$(1) \qquad k(M_k - m_k) < 1 \quad \text{for all} \quad k \in \mathbb{N}.$$

Indeed, let $x_1, x_2 \in \mathbb{R}$ be arbitrary points. Since ψ_k is $\mathbb{Z}$-periodic, we may assume $x_2 < x_1 < x_2 + 1$. Then f increasing implies

$$\psi_k(x_2) - \psi_k(x_1) = f^k(x_2) - f^k(x_1) + x_1 - x_2 < 1 ,$$

so

$$\max_x \psi_k(x) < \min_x \psi_k(x) + 1 .$$

For $\ell \in \mathbb{N}$, we obtain

$$f^{\ell k}(x) = f^k(f^{(\ell-1)k}(x)) = f^{(\ell-1)k}(x) + \psi_k(f^{(\ell-1)k}(x))$$

and therefore

(2) $$k\, m_k \leq f^{\ell k}(x) - f^{(\ell-1)k}(x) \leq k\, M_k, \quad \text{for all } \ell \in \mathbb{N},\ x \in \mathbb{R}.$$

Taking $k = 1$ yields

(3) $$m_1 \leq f^{\ell}(x) - f^{\ell-1}(x) \leq M_1, \quad \text{for all } \ell \in \mathbb{N},\ x \in \mathbb{R}.$$

Let $n,k \in \mathbb{N}$ and write $n = kq + r$, with $0 \leq r < k$. Summing the inequality (2) over ℓ from 1 to q and the inequality (3) over ℓ from $kq + 1$ to $kq + r$ gives

$$kqm_k + rm_1 \leq f^n(x) - x \leq kq\, M_k + r\, M_1, \quad \text{for all } x \in \mathbb{R},$$

and therefore

(4) $$kqm_k + rm_1 \leq nm_n \leq nM_n \leq kqM_k + r\,M_1 .$$

Dividing (4) by n and taking $n \to \infty$ $(n = kq + r,\ 0 \leq r < k)$ yields

(5) $$m_k \leq \liminf_n m_n \leq \limsup_n M_n \leq M_k.$$

Looking at (1) we notice that $M_k - m_k \to 0$ as $k \to \infty$. Thus

(6) $$\liminf_n m_n = \limsup_n M_n,$$

and we denote this limit by $\rho(f)$.
Applying (6), we obtain

$$\rho(f) = \lim_{n\to\infty} \frac{1}{n}\, \psi_n(x) = \lim_{n\to\infty} \frac{(f^n - id)(x)}{n} = \lim_{n\to\infty} \frac{f^n(x)}{n}. \ \square$$

We notice the following additivity property of ρ which in particular implies

$$\rho(f) = \lim_{|n|\to\infty} \frac{f^n(x)}{n} \quad \text{for any} \quad x \in \mathbb{R}.$$

5.1.2. *Lemma*. - Let $f,g \in D^o(S^1)$. If $f \circ g = g \circ f$ then

$$\rho(f \circ g) = \rho(f) + \rho(g).$$

Proof : Assuming $f \circ g = g \circ f$ we obtain

$$\rho(f \circ g) - \rho(g) = \lim_{n\to\infty} \frac{f^n(g^n(x)) - g^n(x)}{n}$$
$$= \lim_{n\to\infty} \frac{\psi_n(g^n(x))}{n} ,$$

with $\psi_n = f^n - id$. We have

$$m_n(f) \leq \frac{\psi_n(g^n(x))}{n} \leq M_n(f), \quad \text{for all} \quad n \in \mathbb{N}, \quad x \in \mathbb{R},$$

and $\lim_{n\to\infty} m_n = \lim_{n\to\infty} M_n = \rho(f)$ thus proving the lemma. □

5.1.3. *Lemma*. - Let $f, g \in D^o(S^1)$ and $h = \psi + id$, where ψ is $\mathbb{Z}$-periodic. If $h \circ f = g \circ h$ then $\rho(f) = \rho(g)$.

Proof (See Herman [He]): For $n \in \mathbb{N}$ we have $h \circ f^n = g^n \circ h$.
Therefore

$$h \circ f^n = f^n + \psi \circ f^n = g^n \circ h + h - h$$

and

$$\frac{f^n - id}{n} + \frac{\psi \circ f^n}{n} = \frac{(g^n - id) \circ h}{n} + \frac{\psi}{n} .$$

For $n \to \infty$ we deduce

$$\rho(f) = \lim_{n\to\infty} \frac{(g^n - id) \circ h}{n} = \rho(g). \ □$$

5.1.4. *Remark.* - In 5.1.3., h need not be a homeomorphism. If however $h \in D^o(S^1)$ then $\rho(f) = \rho(h^{-1} \circ g \circ h) = \rho(g)$. In particular, for $g = R_\alpha$, we get $\rho(f) = \alpha$.

Now consider $\bar{f} \in \mathrm{Diff}_+^o(S^1)$. For any two representatives f_1 and f_2 of $\bar{f}$ one has $\rho(f_1) \equiv \rho(f_2) \mod 1$, according to 5.1.2, with $g = R_n$.

5.1.5. *Definition.* - For $\bar{f} \in \mathrm{Diff}_+^o(S^1)$, the number $\rho(\bar{f}) = \rho(f) \mod 1$, where f is any representative of $\bar{f}$, is called the rotation number of $\bar{f}$.

In particular, $\rho(\bar{R}_\alpha) = \alpha.\ (\mathrm{mod}\ 1)$.

To conclude this section let us record another important property of the rotation number which will be used in 5.3.

5.1.6. *Proposition.* - Let $\bar{f} \in \mathrm{Diff}_+^o(S^1)$ then $\bar{f}^q$ has a fixed point if and only if $q\rho(\bar{f}) \in \mathbb{Z}$.

Proof : Suppose $\bar{f}^q$ has a fixed point $\tilde{x}$, i.e. $f^q(x) = x + p$ for some $p \in \mathbb{Z}$, $e^{2\pi i x} = \tilde{x}$, and $f \in D^o(S^1)$ representing $\bar{f}$. In this case

$$\rho(f) = \lim_{n\to\infty} \frac{f^{nq}(x)}{nq} = \lim_{n\to\infty} \frac{x+np}{nq} = \frac{p}{q} .$$

To show the converse we define

$$a_q = q\,\rho(f) - \min_x \{f^q(x) - x\}$$

and

$$b_q = \max_x \{f^q(x) - x\} - q\,\rho(f).$$

With the notations of 5.1.1. we have

$$m_q \leq \rho(f) \leq M_q, \quad \text{for all } q \in \mathbb{N}.$$

or $$q\, m_q \leq q\, \rho(f) \leq q\, M_q .$$

From this last inequality it follows that $a_q \geq 0$ and $b_q \geq 0$. Therefore

$$\mathrm{im}(f^q - \mathrm{id} - q\, \rho(f)) = [-a_q, b_q]$$

contains 0. Setting $\rho(f) = \frac{p}{q}$ it follows

$$f^q(x) = x + p \quad \text{for some} \quad x \in \mathbb{R}. \square$$

5.2. *Denjoy's example.*

The qualitative behaviour of a suspension F on T^2 obtained by the homeomorphism $f \in \mathrm{Diff}^o_+(S^1)$ will depend heavily on the nature of those subsets of S^1 that are invariant under f. As mentioned before (see p.33), the periodic points of f give rise to compact leaves of F. There is another important phenomenon in this direction, discovered by Denjoy [De] which we are going to describe now. (Recall that a Cantor set in $\mathbb{R}$ (resp. S^1) is a closed subset of $\mathbb{R}$ (resp. S^1) without isolated points and without interior points).

5.2.1. *Denjoy's example* (see 5.2.9 for a more precise statement). There is an orientation preserving C^1 diffeomorphism of S^1 without periodic points which keeps a Cantor set invariant.

5.2.2. *Remark.* - The suspension of such a diffeomorphism admits an exceptional minimal set and thus also exceptional leaves. Foliations with this property have been announced in section 4. This should be contrasted with 5.3.

We shall show the existence of such an example by constructing a representative in $D^1(S^1)$. Roughly speaking, the idea of this construction is a follows. We take a countable dense set $A \subset \mathbb{R}$ and cut $\mathbb{R}$

in the points of A. For each $x \in A$ we fill in an interval where all these intervals have bounded length. We construct a continuous increasing map $h : \mathbb{R} \to \mathbb{R}$ which is the "identity" outside the intervals filled in. This defines a homeomorphism $f \in D^{o}(S^{1})$ such that $h \circ f = R_{\alpha} \circ h$ for some irrational $\alpha \in \mathbb{R}$. The homeomorphism f has no periodic points and keeps a Cantor set invariant. A certain amount of work is necessary to modify f so that it becomes C^{1}.

We now give a precise description which is inspired by Rosenberg's exposition [Ro].

5.2.3. *The cutting process*. This is described by an increasing map $J : \mathbb{R} \to \mathbb{R}$.

We fix $\alpha \in \mathbb{R} - \mathbb{Q}$. By G we denote the subgroup of $\mathrm{Diff}_{+}^{\infty}(\mathbb{R})$ generated by the two translations R_{α} and R_{1}. Each element $g \in G$ can be written in a unique way as $g = R_{\alpha}^{n} \circ R_{1}^{m}$ with $(n,m) \in \mathbb{Z}^{2}$. For $x \in \mathbb{R}$, the G-orbit of x is denoted by $G(x)$, i.e.

$$\begin{aligned} G(x) &= \{y \in \mathbb{R} \mid y = g(x) \quad \text{for some} \quad g \in G\} \\ &= \{y \in \mathbb{R} \mid y = n\alpha + m + x \qquad \text{for some} \quad (n,m) \in \mathbb{Z}^{2}\}. \end{aligned}$$

Finally, let $u_{o} \in \mathbb{R} - G(0)$ and $\ell_{n} > 0$, $n \in \mathbb{Z}$, such that $\sum_{n \in \mathbb{Z}} \ell_{n} = \frac{1}{2}$.

We define a weight function $p : \mathbb{R} \to \mathbb{R}^{+}$ by

$$p(t) = \begin{cases} 0 & \text{if} \quad t \notin G(u_{o}) \\ \ell_{n} & \text{if there is } m \in \mathbb{Z} \text{ such that } t = (R_{\alpha}^{n} \circ R_{1}^{m})(u_{o}). \end{cases}$$

This is used to define $J : \mathbb{R} \to \mathbb{R}$ by

$$J(t) = \begin{cases} \dfrac{t}{2} + \sum\limits_{0<\theta\leq t} p(\theta) & \text{for } t \geq 0 \\[2ex] \dfrac{t}{2} - \sum\limits_{t<\theta\leq 0} p(\theta) & \text{for } t < 0. \end{cases}$$

The map J is everywhere continuous from the right and continuous from the left for points not belonging to $G(u_o)$. For $t \in G(u_o)$, $t = (R_\alpha^n \circ R_1^m)(u_o)$, the jump of J in t equals ℓ_n, see fig. 22.

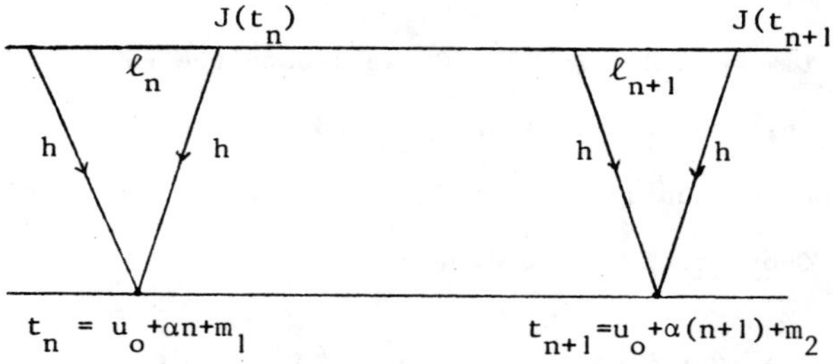

Figure 22

Let C be the closure of $\operatorname{im} J$ in $\mathbb{R}$. Then C is a Cantor set.

5.2.4. *Lemma*. - <u>The map</u> J <u>has the following properties</u> :

(1) <u>There is a (unique) continuous non-decreasing map</u> $h : \mathbb{R} \to \mathbb{R}$ <u>such that</u> $h \circ J = \mathrm{id}_{\mathbb{R}}$. <u>In particular</u> $h(C) = \mathbb{R}$.

(2) $J \circ R_1 = R_1 \circ J$ <u>and</u> C <u>is invariant under</u> R_1.

(3) $h \circ R_1 = R_1 \circ h$.

Proof : Condition (1) is evident, by the construction of J.

To prove (2) we first observe that for every $t \in \mathbb{R}$ and every $n \in \mathbb{Z}$ there exists exactly one $m \in \mathbb{Z}$ such that $(R_\alpha^n \circ R_1^m)(u_o) \in (t,t+1]$. From this we deduce

$$\sum_{t<\theta\leq t+1} p(\theta) = \sum_{n\in\mathbb{Z}} \ell_n = \frac{1}{2} .$$

Consequently, for every $t \geq 0$,

$$J(t+1) = \frac{t}{2} + \frac{1}{2} + \sum_{0\leq\theta\leq t} p(\theta) + \sum_{t<\theta\leq t+1} p(\theta) = J(t) + 1.$$

There is an analogous calculation for $t < 0$. Here the fact that $u_o \notin G(0)$ is used ! It follows from the continuity of R_1 that $C = \mathrm{cl}(\mathrm{im}\, J)$ is also invariant under R_1.

To verify (3) we have to check two cases : First suppose $x = J(t)$. Then, by (2)

$$x + 1 = J(t) + 1 = J(t + 1)$$

and hence $$h(x + 1) = t + 1 = h(x) + 1.$$

By continuity, (3) holds on C.

Now suppose $x \in \mathbb{R} - C$, that is to say $x \in (x_o,y_o)$ where (x_o,y_o) is a component of $\mathbb{R} - C$. Since C is invariant under R_1 the interval $(x_o + 1, y_o + 1)$ is a component of $\mathbb{R} - C$ as well. By the definition of h, we conclude

$$h(x + 1) = h(y + 1) = h(y) + 1 = h(x) + 1. \square$$

5.2.5. *Definition*. - Let $\alpha \in \mathbb{R} - \mathbb{Q}$, $J : \mathbb{R} \to \mathbb{R}$ and $h : \mathbb{R} \to \mathbb{R}$ as in 5.2.3. and 5.2.4. A homeomorphism $f \in D^o(S^1)$ is called a Denjoy homeomorphism (with rotation number α) if $h \circ f = R_\alpha \circ h$, that is f is semi-conjugate to R_α. (Note that $\rho(f) = \alpha$ follows from 5.1.3.).

5.2.6. *Lemma*. - If $f \in D^o(S^1)$ is a Denjoy homeomorphism with $\rho(f) = \alpha$ then for every $x \in \operatorname{im} J$

$$(*) \qquad f(x) = J(h(x) + \alpha).$$

Moreover, $f(C) = C$, where $C = \operatorname{cl}(\operatorname{im} J)$, and C is minimal under the action of the subgroup H of $D^o(S^1)$ which is generated by f and R_1. (This means that no proper closed subset of C is invariant under H).

Proof : First, let $x \in \operatorname{im} J$ such that $h(x) \notin G(u_o)$. Then

$$(1) \qquad h(f(x)) = R_\alpha(h(x)) = h(x) + \alpha \in G(h(x)).$$

By the definition of h we have

$$(2) \qquad J(h(y)) = y \text{ for all } y \in \operatorname{im} J.$$

But $h(f(x)) = h(x) + \alpha \notin G(u_o)$ and hence $f(x) \in \operatorname{im} J$. It follows from (2) and (1) that (*) holds for $x \in \operatorname{im} J$ such that $h(x) \notin G(u_o)$ and that the orbit $H(x)$ is also contained in $\operatorname{im} J$. As $G(h(x))$ is dense in $\mathbb{R}$ it follows from the semi-continuity of J that $H(x)$ is dense in $\operatorname{im} J$ and hence also in C. The continuity of f and h and the semi-continuity of J imply that (*) holds for every $x \in \operatorname{im} J$ and that $f(C) = C$.

Finally, one easily verifies that all orbits $H(x)$, $x \in C$, are dense in C. This proves the lemma. □

Let again C be the Cantor set as above. By $I_{n,m}$ we denote the closure of the component of $\mathbb{R} - C$ which contains the point $J(t)$ as right boundary point, with $t = R_\alpha^n \circ R_1^m(u_o)$. For $f \in \operatorname{Diff}_o^r(\mathbb{R})$ we denote by $f_{n,m}$ the restriction of f to the interval $I_{n,m}$.

5.2.7. *Lemma.* - The map $f \in \mathrm{Diff}^{o}_{+}(R)$ is a Denjoy homeomorphism with $\rho(f) = \alpha$ if and only if

(1) $f(x) = J(h(x) + \alpha)$ for all $x \in \mathrm{im}\, J$, (cf. 5.2.5, 5.2.6).

(2) $f_{n,o} : I_{n,o} \to I_{n+1,o}$ is a homeomorphism for all $n \in \mathbb{Z}$ and

$f_{n,m} = R_m \circ f_{n,o} \circ R_{-m}$ for all $(n,m) \in \mathbb{Z}^2$.

Proof : By 5.2.6., it is clear that every Denjoy homeomorphism satisfies conditions (1) and (2).

To prove the converse we first observe that (1) implies $f(\mathrm{im}\, J) = \mathrm{im}\, J$ and hence $f(C) = C$.

Let $x \in \mathrm{im}\, J$ then, by (1), $f(x+1) = J(h(x+1) + \alpha)$. Therefore, by 5.2.4.,

$$f(x+1) = J(h(x) + 1 + \alpha) = J(h(x) + \alpha) + 1 = f(x) + 1.$$

By continuity we thus have $f \circ R_1 = R_1 \circ f$ on C. But this relation immediately extends to $\mathbb{R}$, by condition (2).

In the same way, condition (1) implies that $h(f(x)) = h(x) + \alpha$ for $x \in \mathrm{im}\, J$. Therefore the relation $h \circ f = R_\alpha \circ h$ holds on C. Since f preserves $\mathbb{R} - C$ this relation extends over $\mathbb{R}$ by the same argument as used at the end of the proof of 5.2.4. □

We are now able to prove :

5.2.8. *Theorem.* - For every irrational $\alpha \in \mathbb{R}$ there exists a Denjoy homeomorphism with rotation number equal to α.

Proof : Let the maps J and h, the Cantor set C and the intervals $I_{n,o}$, $n \in \mathbb{Z}$, be as always in this section. On $I_{n,o}$ we define an increasing homeomorphism $f_{n,o}$ which takes $I_{n,o}$ onto $I_{n+1,o}$, $n \in \mathbb{Z}$. The conditions (1) and (2) of 5.2.7. permit us to extend all these homeomorphisms to a unique homeomorphism $f : \mathbb{R} \to \mathbb{R}$. Then, by construction, f is a Denjoy homeomorphism whose rotation number equals α, by 5.1.3. □

Under certain restrictions on the homeomorphisms $f_{n,o}$ of the preceding theorem we can improve 5.2.8 so that the Denjoy homeomorphism becomes C^1. Recall that for the construction of f we used positive numbers ℓ_n such that $\sum_{n\in\mathbb{Z}} \ell_n = \frac{1}{2}$. (The map $p(\theta)$ which will occur in the proof of 5.2.9 was introduced in 5.2.3.).

5.2.9. *Theorem*. - <u>To every irrational $\alpha \in \mathbb{R}$ there exists a C^1 Denjoy diffeomorphism with rotation number equal to</u> α.

Proof : We choose the positive numbers ℓ_n, $n \in \mathbb{Z}$, in such a way that

$$\sum_{n\in\mathbb{Z}} \ell_n = \frac{1}{2} \quad \text{and} \quad \lim_{|n|\to\infty} \frac{\ell_{n+1}}{\ell_n} = 1.$$

Then the homeomorphisms $f_{n,o}$ used in the proof of 5.2.8. can be chosen to be C^1 diffeomorphisms which, in addition, fulfill the conditions

(1) $Df_{n,o}(x_n) = Df_{n,o}(y_n) = 1$, where $I_{n,o} = [x_n, y_n]$ and $n \in \mathbb{Z}$ is arbitrary,

(2) $\lim_{|n|\to\infty} \sup_{I_{n,o}} |Df_{n,o} - 1| = 0$.

We define $f_{n,m} : I_{n,m} \to I_{n+1,m}$ by the formula $f_{n,m} = R_m \circ f_{n,o} \circ R_{-m}$. Then, by construction, the map $g : \mathbb{R} \to \mathbb{R}$ defined by

$$g(x) = \begin{cases} 1 & \text{for } x \in C \\ Df_{n,m}(x) & \text{for } x \in I_{n,m} \end{cases}$$

is continuous. We claim that the map $f : \mathbb{R} \to \mathbb{R}$ defined by

$$f(x) = \frac{\alpha}{2} + \sum_{0 \leq \theta \leq \alpha} p(\theta) + \int_0^x g(s)ds$$

is a C^1 Denjoy diffeomorphism with rotation number α. To prove this assertion we first observe that f is a C^1 diffeomorphism. It remains therefore to verify the conditions (1) and (2) of 5.2.7.

Let $x > 0$ be a point in $\operatorname{im} J$ and $(n,m) \in \mathbb{Z}^2$ such that $I_{n,m} \subset [0,x]$. Then

$$\int_{I_{n,m}} g(s)ds = \text{length of } I_{n+1,m} = p(R_\alpha(t))$$

$$\text{with } t = (R_\alpha^n \circ R_1^m)(u_o).$$

Consequently

$$\int_0^x g(s)ds = \int_{C \cap [0,x]} ds + \sum_{0 \leq \theta \leq h(x)} p(R_\alpha(\theta)) = \frac{h(x)}{2} + \sum_{0 \leq \theta \leq h(x)} p(R_\alpha(\theta)).$$

Finally,

$$f(x) = \frac{h(x)+\alpha}{2} + \sum_{0 \leq \theta \leq \alpha} p(\theta) + \sum_{0 \leq \theta \leq h(x)} p(R_\alpha(\theta)) = \frac{h(x)+\alpha}{2} + \sum_{0 \leq \theta \leq h(x)+\alpha} p(\theta).$$

By definition of J we get

$$f(x) = J(h(x) + \alpha)$$

which is condition (1) of 5.2.7.

There is a similar calculation for $x < 0$. For $x = 0$ (1) holds trivially.

To establish condition 5.2.7., (2) it suffices to realize that the restriction of f to $I_{n,m}$ coincides with $f_{n,m}$. □

We conclude this section with an immediate consequence of 5.2.9.

The diffeomorphism f constructed above belongs to $D^1(S^1)$. Therefore the equality $h \circ f = R_\alpha \circ h$ projects to S^1 to give $\bar{h} \circ \bar{f} = \bar{R}_\alpha \circ \bar{h}$. Moreover, $\bar{f}$ preserves the Cantor set $C' = \pi(C)$, where $\pi : \mathbb{R} \to S^1$ is the canonical covering map. We have thus the

5.2.10. Theorem. - For every irrational $\alpha \in \mathbb{R}$ there is a C^1 diffeomorphism $\bar{f} : S^1 \to S^1$ such that

(1) $\bar{f}$ is semi-conjugate to $\bar{R}_\alpha$,

(2) there is a Cantor set in S^1 which is invariant under $\bar{f}$.

5.3. *Denjoy's theorem.*

If $f \in \mathrm{Diff}^0_+(S^1)$ then we already know from section 4 that the non-empty minimal closed invariant subsets of S^1 under f (and its powers) can be one of

(1) a finite set,

(2) all of S^1,

(3) a Cantor set, i.e. exceptional minimal.

Denjoy's example is of course an example for (3). The following theorem, also due to Denjoy [De] , excludes the possibility (3) when $f \in \mathrm{Diff}^2_+(S^1)$. This result gives a negative answer to a question of Poincaré [Po] concerning analytic vector fields on T^2. It was the origin of many investigations some of which will be discussed in the sequel, (see .....). Expressed in terms of foliations it will give us first examples which tell us that there is an essential difference in the qualitative behaviour of C^1 and C^2 foliations.

5.3.1. *Theorem*. - Let $\bar{f} \in \mathrm{Diff}^2_+(S^1)$ then $\bar{f}$ has no exceptional minimal set.

Before we begin with the proof of this theorem we state two corollaries.

5.3.2. *Corollary*. - Let $\bar{f} \in \mathrm{Diff}^2_+(S^1)$. If the rotation number of $\bar{f}$ is irrational then all orbits of $\bar{f}$ are dense in S^1.

Proof : We know from 5.1.6. that $\bar{f}$ does not have any periodic orbit. By 5.3.1., case (3) above is excluded so that only possibility (2) remains. □

5.3.3. *Corollary*. - Let $\bar{f} \in \mathrm{Diff}^2_+(S^1)$. If $\alpha = \rho(\bar{f})$ is irrational then $\bar{f}$ is topologically conjugate to the rotation $\bar{R}_\alpha$.

Proof : Let $f \in D^2(S^1)$ be a representative of $\bar{f}$ and let G be the subgroup of $D^2(S^1)$ generated by f and R_1. Then, since $\alpha \in \mathbb{R} - \mathbb{Q}$ no element of G has a fixed point. Therefore we get a total order on G by

$g \leq h$ if and only if $g(x) \leq h(x)$ for (one and therefore) all $x \in \mathbb{R}$.

Furthermore, G provided with this relation has the Archimedean property that is if $g > \mathrm{id}$ then for every $h \in G$ there is $n \in \mathbb{N}$ such that $g^n > h$.

We can therefore apply Hölder's theorem (see Birkhoff [Bi]) to get an order preserving group monomorphism

$$\emptyset : (G,\circ) \to (\mathbb{R},+).$$

Clearly, we may suppose $\emptyset(R_1) = 1$.

For $u_o \in \mathbb{R}$, let $G(u_o)$ be the orbit of u_o under G.

We define

$$\lambda : G(u_o) \to G \quad \text{by} \quad \lambda(g(u_o)) = g.$$

The map

$$\psi = \vartheta \circ \lambda : G(u_o) \to \mathbb{R}$$

is increasing and $G(u_o)$ and $\psi(G(u_o))$ are both dense in $\mathbb{R}$. It follows that there is a unique extension $\Psi : \mathbb{R} \to \mathbb{R}$ of ψ and that Ψ is a homeomorphism.

We thus have

$$\Psi \circ g = \vartheta(g) \circ \Psi \,, \quad \text{for all} \quad g \in G,$$

in particular $\Psi \circ f \circ \Psi^{-1} = R_\beta$, where $\beta = \vartheta(f)$. An easy calculation shows that $\psi(x+1) = \psi(x)+1$, so the same must hold for Ψ. Consequently

$$\bar{\Psi} \circ \bar{f} \circ \bar{\Psi}^{-1} = \bar{R}_\beta .$$

By 5.1.3., we conclude $\beta = \alpha \mod 1$ and hence $\bar{R}_\beta = \bar{R}_\alpha$. □

Remark. An alternative proof of 5.3.3. using invariant measures can be found in Herman [He].

We now come to the proof of 5.3.1. which will be carried out in several steps. Our method of proof is inspired by Schwartz [Sc]. A slightly different proof can be found in Siegel [Sie]. It was probably Schwartz's work that inspired also Sacksteder for the proof of his result [Sa]. Our proof of Sacksteder's theorem in chapter VI will be a somewhat more elaborate version of the following.

First some notations. Let $\bar{f} \in \mathrm{Diff}^o_+(S^1)$ and let M be an exceptional minimal set of $\bar{f}$. A component J of $S^1 - M$ is of the form $J = \pi((x,y))$ where $\pi : \mathbb{R} \to S^1$ is the canonical covering projection and (x,y) is an open interval. Therefore we can write $J = (s,t)$ with $s = \pi(x)$, $t = \pi(y)$. The component J as well as its closure, which is denoted by $[s,t]$, are also called intervals. The length of J is by definition the length of $[x,y]$. (This is obviously well defined).

If $\bar{f} \in \mathrm{Diff}^0_+(S^1)$ then $f \in D^0(S^1)$ is a representative of $\bar{f}$ which is fixed once and for all. As always, f^n (resp. $\bar{f}^n$) denotes the n-th iterate of f (resp. $\bar{f}$).

5.3.4. *Lemma*. - Assume $\bar{f} \in \mathrm{Diff}^0_+(S^1)$ has an exceptional minimal set M. Let (s_o,t_o) be a component of $S^1 - M$, let $I_o = [s_o,t_o]$ and let ℓ_n be the length of $I_n = \bar{f}^n(I_o)$. Then

(1) $\bar{f}$ has no periodic point,

(2) the intervals I_n, $n \in \mathbb{Z}_+$, are mutually disjoint,

(3) $\sum_{n\in\mathbb{Z}_+} \ell_n \leq 1$, in particular $\lim_{n\to\infty} \ell_n = 0$

Proof : (3) follows from (2) and (2) from (1), so we have to prove (1).

Let $t_o \in S^1$ such that $\bar{f}^n(t_o) = t_o$ for some $n \in \mathbb{N}$. As M is minimal every orbit contained in M is dense in M. Therefore $t_o \in S^1 - M$ and there is a component $(s,t) \subset S^1 - M$ which contains t_o. Hence $\bar{f}^n([s,t]) = [s,t]$ and thus $\bar{f}^n(s) = s$ which is impossible (M cannot contain a compact leaf). □

5. 3.5. - *Observation*. - For $f \in D^2(S^1)$ we denote by $D^k f$ $(k = 1,2)$ its k-th derivative. There exists $\theta > 0$ such that

$$|D^2 f(t)| \leq \theta\, Df(t) \text{ for every } t \in S^1 .$$

5.3.6. *Lemma*. - Suppose $f \in D^2(S^1)$ and let $[x,y]$ be an interval in $\mathbb{R}$. Then for every $n \in \mathbb{N}$ one has

$$\left|\log \frac{Df^n(x)}{Df^n(y)}\right| \leq \theta \sum_{j=0}^{n-1} |f^j(y) - f^j(x)|.$$

Proof : For every $n \in \mathbb{N}$ we have

$$Df^n(x) = \prod_{j=0}^{n-1} Df(f^j(x)).$$

Therefore

$$\log \frac{Df^n(x)}{Df^n(y)} = \sum_{j=0}^{n-1} (\log Df(f^j(x)) - \log Df(f^j(y))).$$

By the mean value theorem, there are $z_j \in [x,y]$ such that

$$\log \frac{Df^n(x)}{Df^n(y)} = \sum_{j=0}^{n-1} \frac{D^2f(z_j)}{Df(z_j)} (f^j(x) - f^j(y)).$$

The lemma follows from 5.3.5. □

5.3.7. *Lemma*. - Suppose that $\bar{f} \in \mathrm{Diff}^2_+(S^1)$ has an exceptional minimal set M. Let $J = \pi((x_o,y_o))$ be a component of $S^1 - M$ and let ℓ_n be the length of $\bar{f}^n(J)$. Then for every $n \in \mathbb{N}$ and every $x \in [x_o,y_o]$ we have

$$Df^n(x) \leq e^\theta \frac{\ell_n}{\ell_o}.$$

Proof : Let $x,y \in [x_o,y_o]$. From 5.3.6 and 5.3.4., (3) we deduce

$$\left|\log \frac{Df^n(x)}{Df^n(y)}\right| \leq \theta \sum_{j=0}^{n-1} \ell_j \leq \theta.$$

Hence

$$Df^n(x) \leq e^\theta Df^n(y).$$

The mean value theorem provides $z_n \in [x_o, y_o]$ such that $\ell_n = \ell_o \; Df^n(z_n)$. For $y = z_n$ the last inequality becomes

$$Df^n(x) \leq e^{\theta} \frac{\ell_n}{\ell_o} . \square$$

Lemma 5.3.7. means that $\{Df^n\}$, $n \in \mathbb{N}$ converges uniformly to zero on $[x_o, y_o]$. We want to show that this uniform convergence also holds in a full neighbourhood of x_o.

We set

$$\nu = \frac{\ell_o}{\theta e^{\theta+1}} \quad \text{and} \quad U = [x_o - \nu, y_o] .$$

5.3.8. *Lemma*. - With the same hypotheses as in 5.3.7. we have for every $n \in \mathbb{Z}_+$ and every $x \in U$ with $x < x_o$

$$Df^n(x) \leq e \; Df^n(x_o).$$

Proof : The inequality is obviously true for $n = 0$ so let us assume that it holds for all j, $0 \leq j \leq n-1$.

By 5.3.6., we have for every $n \in \mathbb{N}$ and every $x \in U$, $x < x_o$,

$$\left| \log \frac{Df^n(x)}{Df^n(x_o)} \right| \leq \theta \sum_{j=0}^{n-1} |f^j(x) - f^j(x_o)|.$$

The mean value theorem yields numbers $z_j \in [x, x_o]$ such that

$$\left| \log \frac{Df^n(x)}{Df^n(x_o)} \right| \leq \theta \; \nu \sum_{j=0}^{n-1} Df^j(z_j).$$

By the induction hypothesis we conclude

$$\left| \log \frac{Df^n(x)}{Df^n(x_o)} \right| \leq \theta \; \nu \; e \sum_{j=0}^{n-1} Df^j(x_o).$$

Finally, using 5.3.7., 5.3.4., (3) and the special choice of ν we get

$$\left|\log \frac{Df^n(x)}{Df^n(x_o)}\right| \leq \theta\, \nu\, e\, e^{\theta} \sum_{j=o}^{n-1} \frac{\ell_j}{\ell_o} \leq \frac{\theta\, \nu\, e^{\theta+1}}{\ell_o} = 1$$

and the lemma follows. □

Combining the last two lemmas we immediately get the required convergence on U :

5.3.9. *Lemma*. - Under the hypotheses of 5.3.7., we get for every $n \in \mathbb{N}$ and every $x \in U$

$$Df^n(x) \leq e^{\theta+1} \frac{\ell_n}{\ell_o} .$$

We now come to the proof of Denjoy's theorem as it was announced at the beginning of this section. We use the same notations as before.

Proof of 5.3.1. : Since $s_o = \pi(x_o) \in M$, the orbit $\{\bar{f}^n(s_o)\}$, $n \in \mathbb{Z}$, is dense in M. Therefore, there exists a sequence $\{\varphi(k)\}$, $k \in \mathbb{N}$, in $\mathbb{Z}$ such that

$$s_o = \lim_{k\to\infty} \bar{f}^{\varphi(k)}(s_o) .$$

Possibly after replacing $\bar{f}$ by $\bar{f}^{-1}$ we may assume that all $\varphi(k)$ are positive.

By 5.3.9., there is $k \in \mathbb{N}$ such that for $j = \varphi(k)$ one has

(1) $Df^j(x) < \frac{1}{2}$ for all $x \in U$,

(2) $\bar{f}^j(s_o) \in \pi(U_\nu)$, where $U_\nu = [x_o - \frac{\nu}{2}, y_o]$.

It follows from (1) and (2) that $\bar{f}^j(V) \subset V$, where $V = \pi(U)$. Iterating $\bar{f}^j$ yields a fixed point for $\bar{f}^j$, that is a periodic point of $\bar{f}$. This contradicts 5.3.4., (1). □

6. *Structural stability.*

The notion of structural stability has its origin in the study of mechanical systems. If a motion is described by a vector field then one wants to know whether its family of solution curves is "structurally stable", i.e. whether the qualitative behaviour is unchanged when the vector field is replaced by one which is nearby.

In terms of foliations the problem consists first in choosing a suitable topology for the set of foliations under consideration, (we shall be interested in foliations of class C^1 admitting a C^1 tangent vector field, cf. 6.2.1., i), and then asking which are the structurally stable foliations with respect to this topology. Such a foliation is required to possess a neighbourhood all of whose elements are pairwise topologically conjugate ; see 6.2.1.

It should be pointed out that, in contrast to that for higher dimensional foliations, the problem of structural stability for foliations on compact surfaces is particularly simple. Indeed, one gets the two following basic results :

i) The structurally stable foliations are characterized by the fact that they have at least one compact leaf and all compact (circle) leaves have non-trivial linear holonomy ; see 6.3.12.

ii) The subset of structurally stable foliations turns out to be open and dense; see 6.3.13.

We shall be interested mainly in foliations on the torus but in order to handle those it is necessary first to study structurally stable foliations on the annulus.

For simplicity we restrict ourselves to the study of orientable foliations. On the annulus this is no restriction at all, (cf. 2.3.11., 4.2.15.). Non-orientable foliations as well as foliations on the Möbius band and foliations on the Klein bottle are treated in exercices 6.2.12. and 6.3.14.

6.1. *Structural stability for diffeomorphisms of the interval and the circle.*

We already know from several of our previous considerations that there is a close relationship between foliations on compact surfaces and diffeomorphisms of the interval and the circle. As we shall see, such a relationship exists also with regard to structural stability. Therefore we begin by investigating the structural stability for diffeomorphisms of the interval and the circle. Since this matter seems to be fairly well known our style in this section is more concise than in other parts of this chapter. Proofs will sometimes only be sketched. The reader unfamiliar with the subject may consult the literature, for instance Nitecki [Ni].

6.1.1. *Definitions and remarks.*- For $K = I$ or S^1 we denote by $\mathrm{Diff}^1(K)$ the group of C^1 diffeomorphisms of K endowed with the topology of C^1 uniform convergence. Clearly, the subgroup $\mathrm{Diff}^1_+(K)$ of orientation preserving elements is an open subset.

i) The diffeomorphisms f and g of $\mathrm{Diff}^1(K)$ are (strongly) conjugate if there is an orientation preserving homeomorphism h of K such that $g = h^{-1}fh$.

Evidently, this defines an equivalence relation on $\mathrm{Diff}^1(K)$ whose corresponding equivalence classes are called the conjugacy classes of $\mathrm{Diff}^1(K)$.

ii) For the characterization of the conjugacy classes we shall use the sets Fix(f) (resp. Per(f)) of fixed points (resp. periodic points) of $f \in \mathrm{Diff}^1(K)$. Both Fix(f) and Per(f) are closed subsets of K.

If f and g are conjugate by the homeomorphism h then $h(\mathrm{Fix}(g)) = \mathrm{Fix}(f)$ and $h(\mathrm{Per}(g)) = \mathrm{Per}(f)$.

We say in this case that the <u>fixed points</u> (resp. <u>the periodic points</u>) of f and g <u>are conjugate</u> by h.

(iii) The element $f \in \mathrm{Diff}^1(K)$ is called <u>structurally stable</u> (or simply <u>stable</u>) if there exists a neighbourhood W of f in $\mathrm{Diff}^1(K)$ such that every $g \in W$ is conjugate to f.

For $f \in \mathrm{Diff}^1_+(K)$ it suffices to find a neighbourhood W as above in $\mathrm{Diff}^1_+(K)$ to assure that f is stable.

(iv) If f is stable then, for every g sufficiently close to f, Fix(g) and Per(g) are conjugate to Fix(f) and Per(f), respectively, thus providing a necessary condition for f being stable. We say in this case that Fix(f) (resp. Per(f)) is <u>stable</u>.

We are now going to characterize the structurally stable C^1 diffeomorphisms of K. Here we consider only orientation preserving maps. The orientation reversing diffeomorphisms are treated in exercice 6.1.11.

A) Let us begin with $K = \bar{I}$. Observing that lemma 4.2.7. has a trivial converse we get a first special result :

6.1.2. - *Lemma*.- Let $f \in \mathrm{Diff}_+^1(I)$ be above (resp. below) the diagonal, (cf. 4.2.6.). Then f is stable if and only if it admits a neighbourhood all of whose elements are above (resp. below) the diagonal.

6.1.3. - *Definition*.- A fixed point x of $f \in \mathrm{Diff}_+^1(I)$ is called hyperbolic if $Df(x) \neq 1$. The set $\mathrm{Fix}(f)$ is hyperbolic if each $x \in \mathrm{Fix}(f)$ is hyperbolic. In this case we say that f is hyperbolic.

The next result is a standard fact. It is obtained by looking at the graphs of the diffeomorphisms under consideration.

6.1.4. - *Proposition*.- The subset of hyperbolic elements is open and dense in $\mathrm{Diff}_+^1(I)$.

Using 6.1.4. our first central result of this paragraph can be proved.

6.1.5. - *Theorem*.- An element $f \in \mathrm{Diff}_+^1(I)$ is structurally stable if and only if it is hyperbolic.

Proof : Assume that f is stable but has a non-hyperbolic fixed point. It is not hard to see that then f can be approximated by elements with infinitely many fixed points.

On the other hand, by 6.1.4., there are hyperbolic elements arbitrarily close to f, which clearly have only finitely many fixed points. We deduce that f must be unstable, contradicting our assumption.

Now let f be hyperbolic and $0 = x_o < x_1 < \ldots < x_k = 1$ its fixed points. By looking at the graph of f, we can see that every g sufficiently close to f has the same number of fixed points $0 = y_o < y_1 < \ldots < y_k = 1$ as f and the restrictions of f to the interval $[x_i, x_{i+1}]$ and of g to the interval $[y_i, y_{i+1}]$ are either both above or both below the diagonal. By proceeding as in 4.2.7. for each interval, we construct a homeomorphism h such that $h(x_i) = y_i$, $i = 0,\ldots,k$, and $f = h^{-1}gh$. □

6.1.6. - Remark. - If $f \in \mathrm{Diff}^1_+(I)$ is non-hyperbolic then it can be approximated by a sequence $\{f_n\}$ such that for each n

(1) $\mathrm{Fix}(f_n)$ is finite if $\mathrm{Fix}(f)$ is infinite, and vice versa.

(2) $Df_n(i) = Df(i)$ for $i = 0,1$.

The following exercises are to clarify the preceding discussion.

6.1.7. - Exercises. i) Construct two diffeomorphisms $f,g \in \mathrm{Diff}^1_+(I)$ such that $\mathrm{Fix}(f) = \mathrm{Fix}(g) = \{0,1\}$ which are not conjugate.

ii) Call two elements $f,g \in \mathrm{Diff}^1_+(K)$ (strongly) C^1-conjugate if there exists $h \in \mathrm{Diff}^1_+(K)$ $(K = I$ or $S^1)$ such that $f = h^{-1}gh$.

Show that (strong) C^1-conjugacy is a finer relation than (strong) conjugacy.

Define the notion of C^1 structural stability and show that no element of $\mathrm{Diff}^1_+(K)$ is C^1 structurally stable.

B) We now turn to self-diffeomorphisms of the circle. A first observation is that the rôle of fixed points in the previous study of stability in $\mathrm{Diff}^1_+(I)$ is undertaken by the periodic points.

6.1.8. - *Remarks and definitions*.- i) Recall that for $\bar{f} \in \mathrm{Diff}^1_+(S^1)$ the set $\mathrm{Per}(\bar{f})$ is non-empty if and only if there exists $q \in \mathbb{Z}$ such that $q\rho(\bar{f}) \in \mathbb{Z}$, where $\rho(\bar{f})$ is the rotation number of $\bar{f}$. Moreover, if $\mathrm{Per}(\bar{f}) \neq \emptyset$ then all periodic points have the same order, (cf. 5.1.6.). (The order of a periodic point t of $\bar{f}$ is the least positive q such that $\bar{f}^q(t) = t$).

Furthermore, 5.1.3. in connection with 5.1.4. tells us that the rotation number is invariant under conjugation.

ii) The periodic point t of $\bar{f} \in \mathrm{Diff}^1_+(S^1)$ is called hyperbolic if $Df^q(x) \neq 1$ for some (and therefore for any) $x \in \mathbb{R}$ over t, where q is the order of t and $f \in D^1(S^1)$ is any representative of $\bar{f}$; see 3.1.1. We call $\mathrm{Per}(\bar{f})$ and thus $\bar{f}$ hyperbolic if $\mathrm{Per}(\bar{f})$ is non-empty and consists only of hyperbolic points.

The main step in proving the analogues of 6.1.4. and 6.1.5. requires a weak version of a result which can be found in the literature under the name "Closing lemma" ; see Pugh [Pu].

6.1.9.- *Proposition*.- Every element $\bar{f} \in \mathrm{Diff}^1_+(S^1)$ without periodic points is structurally unstable. More precisely, $\bar{f}$ can be approximated by a sequence $\{\bar{f}_n\}$ such that $\mathrm{Per}(\bar{f}_n) \neq \emptyset$ for each n.

Proof : With the notations of 3.1.1, let $f \in D^1(S^1)$ be a representative of $\bar{f}$. We endow $D^1(S^1)$ with the topology of C^1 uniform convergence. Then the curve

$$c : [0,1] \to D^1(S^1)$$
$$\alpha \mapsto R_\alpha \circ f$$

starts at f and in order to prove the proposition it suffices to show that for every $\varepsilon > 0$ there exists $0 < \beta < \varepsilon$ such that

$\rho(R_\beta \circ f)$ is rational.

We fix $0 < \varepsilon \lesssim 1$. If $\bar{R}_\varepsilon \circ \bar{f}$ does not have a periodic point then $\bar{R}_\varepsilon \circ \bar{f}$ has a minimal set M which is either all of S^1 or a Cantor set. (Minimal sets of $\bar{f}$ are defined similar to those of foliations ; also their classification is similar to that for foliations ; see also 4.1.2.). In both cases there is $t \in M$ which is not a boundary point of one of the components of S^1-M. In other words, if $x \in \mathbb{R}$ covers t, then there are integers n and q such that

$$f^n(x) < x + q \lesssim f^n(x) + \varepsilon.$$

Now for fixed x, n and q, we consider the function

$$A : [0,\varepsilon] \longrightarrow \mathbb{R}$$
$$\alpha \longmapsto (R_\alpha \circ f)^n(x).$$

We have

$$A(0) = f^n(x)$$

and

$$A(\varepsilon) = (R_\varepsilon \circ f)^n(x) \geq (R_\varepsilon \circ f^n)(x) = f^n(x) + \varepsilon.$$

It follows from the continuity of A that there is a $\beta \in [0,\varepsilon]$ with $A(\beta) = x+q$. Hence $\rho((R_\beta \circ f)^n)$ is rational and thus also is $\rho(R_\beta \circ f)$. □

The following theorem is the analogue of 6.1.4. and 6.1.5. for the circle. Its proof consists essentially of an application of 6.1.9. and of arguments similar to those in the proof of 6.1.5. where, this time, the periodic points assume the rôle which was played there by the fixed points.

6.1.10.- Theorem.- i) The subset of hyperbolic elements is open and dense in $\mathrm{Diff}^1_+(S^1)$.

ii) An element of $\mathrm{Diff}^1_+(S^1)$ is structurally stable if and only if it is hyperbolic.

In the form of an exercise we treat the structural stability of orientation reversing C^1 diffeomorphisms of $K = I$ or S^1 :

6.1.11.- Exercises.- i) Show that an orientation reversing homeomorphism of I may have periodic points of order two but cannot have periodic points of order greater than two.

ii) Show that every orientation reversing homeomorphism of K has exactly one fixed point when $K = I$ and exactly two fixed points when $K = S^1$.

iii) If $\bar{f} \in \text{Homeo}(S^1)$ reverses orientation then $\text{Per}(\bar{f}) = \text{Per}(\bar{f}^2)$ and the two fixed points provided by ii) are the only periodic points of odd order.

iv) If $f \in \text{Diff}^1(K)$ reverses orientation then so does every nearby g.

v) Define the notions of structural stability and hyperbolicity for orientation reversing C^1 diffeomorphisms of K.

vi) Prove that the results 6.1.4., 6.1.5., 6.1.10. remain valid when $\text{Diff}^1_+(K)$ is replaced by $\text{Diff}^1(K)$.

6.2. *Structural stability for suspensions.*

Originally, the notion of structural stability was introduced for the study of the topological behaviour of vector fields under small perturbation of the initial values. We want to present an important part of this study in terms of foliations. This will be done using the results of section 6.1.

For simplicity we treat only orientable (and thus, by 2.3.11., also transversely orientable) foliations on the annulus and on the torus. The case where the surface is non-orientable is discussed in exercise 6.2.12. ii).

Our first task is to find a suitable topology for the set of foliations under consideration.

6.2.1. - *Remarks and definitions*.- Let $\Sigma = S^1 \times I$ or $S^1 \times S^1$.

i) By 2.3.2., every orientable C^1 foliation on Σ is defined by a vector field. We are interested in the set of foliations $\mathcal{F}^1(\Sigma)$ where this vector field is of class C^1. Thus $\mathcal{F}^1(\Sigma)$ is a genuine subset of the set of all orientable C^1 foliations on Σ which is sometimes also denoted by $\mathcal{F}^1(\Sigma)$ in the literature. Of course, if $\mathcal{F} \in \mathcal{F}^1(\Sigma)$ is defined by the vector field X and f is any nowhere vanishing C^1 function on Σ then $\mathcal{F}$ is also defined by fX.

Let $\mathcal{X}^1(\Sigma)$ be the set of C^1 vector fields without singularities on Σ which on each boundary component are either transverse or tangent to $\partial\Sigma$, when $\partial\Sigma \neq \emptyset$. ("Transversality" now means "C^1 transversality"). The equivalence relation ρ on $\mathcal{X}^1(\Sigma)$ given by

$$X \,\rho\, Y \text{ if } Y = fX \text{ for some } C^1 \text{ function } f : \Sigma \to \mathbb{R} - \{0\}$$

yields a quotient map

$$\pi : \mathcal{X}^1(\Sigma) \to \mathcal{F}^1(\Sigma).$$

Thus when $\mathcal{X}^1(\Sigma)$ is endowed with the topology of C^1 uniform convergence then π induces a topology on $\mathcal{F}^1(\Sigma)$ which is referred to as the C^1 topology on $\mathcal{F}^1(\Sigma)$.

(ii) The foliations $\mathcal{F}, \mathcal{F}' \in \mathcal{F}^1(\Sigma)$ are called (strongly) C^r-conjugate if there is a C^r diffeomorphism h on Σ which is isotopic to the identity and with $h^*\mathcal{F}' = \mathcal{F}$.

iii) We call $F \in F^1(\Sigma)$ (structurally) stable if there exists a neighbourhood W of F in $F^1(\Sigma)$ such that each $F' \in W$ is (topologically) conjugate to F.

iv) For $F \in F^1(\Sigma)$ we denote by $C(F)$ the set of circle leaves of F. (Sometimes we consider $C(F)$ as a subset of Σ). If the foliations F and F' are conjugate then $C(F)$ and $C(F')$ are conjugate in a sense similar to 6.1.1., ii). This gives a necessary condition for the stability of F.

We first consider structural stability for a special subset of $F^1(\Sigma)$ before treating the general case in the next section.

Let ξ be the product bundle

$$p : S^1 \times K \to S^1.$$

(We take the product bundle only for convenience. Any other bundle would do just as well). Denote by $\frac{\partial}{\partial t}$ the canonical vector field on S^1.

6.2.2.- Definitions and remarks.- i) The vector field X on $\Sigma = S^1 \times K$ is projectable on S^1 if

$$Tp \circ X = \frac{\partial}{\partial t} \circ p,$$

where Tp denotes the differential of p.

ii) We denote by $X^1(\xi)$ the subspace of $X^1(\Sigma)$ which consists of the projectable elements. The subspace of $F^1(\Sigma)$ corresponding to $X^1(\xi)$ is denoted by $F^1(\xi)$.

iii) Note that $F^1(\xi)$ is exactly the subset of $F^1(\Sigma)$ consisting of the foliations which are transverse to the fibres of ξ. Moreover, for each $F \in F^1(\xi)$ there is a unique projectable vector field X

defining F, and the topology on $F^1(\xi)$ as a subspace of $F^1(\Sigma)$ coincides with the topology which is induced from $X^1(\xi)$ by identification of X with F.

iv) A vector field $X \in F^1(\xi)$ generates a flow ϕ_t on Σ and by definition of projectable vector fields, the time-one map ϕ_1 preserves each fibre of ξ.

6.2.3.- *Definition*.- If K is identified with the fibre over $1 \in S^1$ then the map ϕ_1 gives a map

$$f : K \to K$$

which is called the first-return map of X (or of F, the foliation corresponding to X). In the literature f is also called the time-one-map or the Poincaré map of X.

In order to establish the connection between foliations and diffeomorphisms we have to restrict ourselves to yet another subset of $F^1(\Sigma)$ which is even smaller than $F^1(\xi)$ when $\partial\Sigma \neq \emptyset$, namely the subset $F^1_\partial(\xi)$ of $F^1(\xi)$ consisting of the foliations which are tangent to the boundary. Of course $F^1_\partial(\xi)$ is not open in $F^1(\xi)$ when $\partial\Sigma \neq \emptyset$.

Note that, for $F \in F^1_\partial(\xi)$, the first-return map f is a C^1 diffeomorphism of K and that in this case there is a 1-1 correspondence between $C(F)$ and the periodic orbits of f.

The suspension of $f \in \mathrm{Diff}^1_+(K)$ belongs to $F^1_\partial(\xi)$ and its first return map equals f, (up to conjugacy and after suitable trivialization of Σ_f).

6.2.4.- *Lemma*.- The map

$$\eta : F^1_\partial(\xi) \rightarrow \mathrm{Diff}^1_+(K)$$

which assigns to $F \in F^1_\partial(\xi)$ its first-return map is continuous.

Proof : Let X^1 be the Banach space of C^1 vector fields on Σ. On a suitable subset Δ of $\mathbb{R} \times \Sigma \times X^1$ we can define the map

$$\phi : \Delta \rightarrow \Sigma$$

by $\phi(t,x,X) = \varphi_t(x)$, where φ_t is the flow generated by X. Now the theory of differential equations on Banach spaces tells us that ϕ is of class C^1, since it can be considered as the flow which is generated by the C^1 vector field V on $\Sigma \times X^1$ defined by $V(x,X) = (0,X(x))$; see Lang [La, p. 131 f.]. This implies the continuity of η. □

Of course, η is not injective, but as the next lemma shows it is surjective and admits a section.

6.2.5.- *Lemma*.- There is a continuous map

$$\sigma : \mathrm{Diff}^1_+(K) \rightarrow F^1_\partial(\xi)$$

such that $\eta \circ \sigma = \mathrm{id}$.

Proof : Given $f \in \mathrm{Diff}^1_+(K)$, we first construct a C^1 vector field on $\hat{\Sigma} = I \times K$ in the following way.

Let

$$\lambda : I \rightarrow I$$

be an increasing C^∞ map which equals 0 in a neighbourhood of 0 and 1 in a neighbourhood of 1. Then the map

$$\Psi_f : I \times K \to \hat{\Sigma}$$

$$(t,x) \mapsto (t,(1-\lambda(t))x + \lambda(t)f(x))$$

is a C^1 diffeomorphism. Let $Z = \frac{\partial}{\partial t}$ be the unit horizontal vector field on $I \times K$ and let

$$\pi : \hat{\Sigma} \to \Sigma$$

be the map which identifies $(0,x)$ with $(1,x)$. Then the vector field $T\Psi_f \circ Z \circ \Psi_f^{-1}$ on $\hat{\Sigma}$ projects to give a C^1 vector field $X_f \in F^1_\partial(\xi)$. If F_f is the foliation defined by X_f then $\sigma(f)$ is defined to be F_f. □

6.2.6.- *Lemma*.- Let f_i be the first-return map of $F_i \in F^1_\partial(\xi)$, $i = 1,2$. If, for $r = 0$ or $r = 1$, f_1 and f_2 are strongly C^r-conjugate, then F_1 and F_2 are strongly conjugate by a C^r diffeomorphism which preserves each fibre of ξ.

Proof : As in 6.2.5. we think of Σ as obtained from $\hat{\Sigma} = I \times K$ by identification of $(0,x)$ with $(1,x)$. Let X_i be the projectable vector field tangent to F_i and let $\hat{X}_i$ be the lift of X_i to $\hat{\Sigma}$, $i = 1,2$. Denote by $\hat{\phi}^i_t$ the flow generated by $\hat{X}_i$. Then the maps

$$F_i : I \times K \to \hat{\Sigma}$$

$$(t,x) \to \hat{\phi}^i_t(0,x)$$

are such that $\hat{F} = F_2 \circ F_1^{-1} : \hat{\Sigma} \to \hat{\Sigma}$ is a C^r diffeomorphism which is compatible with the identification map

$$\pi : \hat{\Sigma} \to \Sigma .$$

This yields an induced foliation preserving C^r diffeomorphism

$$F : (\Sigma, F_1) \rightarrow (\Sigma, F_2)$$

which preserves each fibre of ξ. □

A foliation on $S^1 \times I$ which is tangent to the boundary can be approximated by foliations which are transverse to the boundary. Thus the elements of $F^1_\partial(\xi)$ are all unstable in the sense of 6.2.1., iii). However, we can consider quite naturally the structural stability for elements of $F^1_\partial(\xi)$ merely with respect to neighbourhoods in $F^1_\partial(\xi)$ rather than in $F^1(\xi)$. Clearly, for $\Sigma = T^2$ the two notions agree.

It then turns out that the study of structural stability for foliations of $F^1_\partial(\xi)$ is reduced to the corresponding problem for elements of $\mathrm{Diff}^1_+(K)$.

6.2.7.- Theorem.- <u>A foliation F of $F^1_\partial(\xi)$ is structurally stable in $F^1_\partial(\xi)$ if and only if its first-return map is structurally stable.</u>

Proof : Assume the first-return map f of F is stable. Then choose a neighbourhood W of f in $\mathrm{Diff}^1_+(K)$ such that each $g \in W$ is conjugate to f. It follows by 6.2.6. that each $F' \in \eta^{-1}(W)$ is conjugate to F and thus F is stable.

Now assume that F is stable but its first-return map f is unstable. According to 6.1.4. and 6.1.10., i) there exists a sequence $\{f_n\}$ in $\mathrm{Diff}^1_+(K)$ converging to f and such that $\mathrm{Per}(f_n)$ is not conjugate to $\mathrm{Per}(f)$ for any n. Applying 6.2.5. to f and $\{f_n\}$, we see that $\sigma(f)$ is unstable. But by 6.2.6., $\sigma(f)$ and F are C^1-conjugate. Hence F is unstable, contradicting our assumption. □

The previous theorem yields also a first general result for foliations of $F^1(T^2)$.

6.2.8.- *Corollary*.- Let $F \in F^1(T^2)$ be such that $C(F)$ is either empty or all of T^2. Then F is unstable.

Proof : Checking the proof of 4.3.2. we see that the same result holds also in the C^1 context. Thus if $C(F) = \emptyset$ or T^2 then F is a suspension, i.e. $F \in F^1(\xi)$. Moreover, if f is the first-return map of F then we have $\mathrm{Per}(f) = \emptyset$ or $\mathrm{Per}(f) = S^1$. In both cases 6.1.10. implies that f is unstable. We deduce from 6.2.7. that F is unstable. □

6.2.9.- *Definition*.- Let $F \in F^1_\partial(\xi)$ and $f = \eta(F)$ its first-return map. The set $C(F)$ is called hyperbolic if $\mathrm{Per}(f)$ is hyperbolic.

There are two further corollaries of the foregoing discussion :

6.2.10.- *Proposition*.- A foliation F of $F^1_\partial(\xi)$ is stable in $F^1_\partial(\xi)$ if and only if $C(F)$ is hyperbolic.

6.2.11.- *Proposition*.- The subset of structurally stable foliations is open and dense in $F^1_\partial(\xi)$.

6.2.12.- *Exercises*.- i) If $F_1, F_2 \in F^1_\partial(\xi)$ are strongly C^r conjugate then they are C^r conjugate by a diffeomorphism which preserves each fibre of ξ, $r = 0$ or $r = 1$.

ii) (Structural stability for suspensions on the Möbius band and on the Klein bottle). Let $\Sigma = S^1 \times_{\mathbb{Z}_2} K$, $K = I$ or S^1, and let ξ be the canonical fibration of Σ over S^1 with fibre K.

a) Define as before the spaces $X^1(\xi)$, $F^1(\xi)$ and $F^1_\partial(\xi)$.

b) Introduce the notions of structural stability for foliations of $F^1(\xi)$ and of $F^1_\partial(\xi)$.

c) Give a characterization of the structurally stable foliations of $F^1_\partial(\xi)$ and show that the subset of stable foliations is open and dense.

6.3. *Structural stability for foliations in general.*

Now we have arrived at the central point of our study, namely to characterize the structurally stable (orientable) foliations of $F^1(S^1 \times I)$ and of $F^1(T^2)$. Non-orientable foliations are treated in exercise 6.3.14.

A) Though we are mainly interested in foliations on the torus, we begin again with foliations on the annulus. These foliations - later to be considered as pieces of foliations on T^2 - will carry all the information necessary to decide whether or not a foliation on T^2 is structurally stable.

We shall be using some results of sections 4.2. in the C^1 context. The reader also should keep in mind that transversality in this context always means C^1 transversality.

In what follows we denote by B_i the boundary component $S^1 \times \{i\}$ of $S^1 \times I$, $i = 0,1$. As usual, we denote by ξ the product fibration of $S^1 \times I$ over S^1. Recall also that $C(F)$ is the set of circle leaves of F.

6.3.1.- Lemma.- Let $F \in F^1(S^1 \times I)$ be such that $C(F) = \emptyset$. Then F is C^1-isotopic to ξ rel B_1, (i.e. there is a C^1 isotopy Φ of $S^1 \times I$ keeping B_1 fixed such that $F = \Phi_1^*(\xi)$).

Proof : It suffices to observe that 4.2.5. holds also in the case of C^1 foliations. The additional condition on the isotopy to keep B_1 fixed can easily be established. □

6.3.2.- *Lemma*.- Let $F \in F^1(S^1 \times I)$ be such that $C(F) = B_o$. Then F is C^1-isotopic rel B_1 to a foliation F of $F^1(\xi)$.

Proof : Let U be a distinguished neighbourhood of $x_o \in B_o$. By 4.2.4., B_o is the only minimal set of F. Hence, if L is the leaf starting from $x_1 \in B_1$ then L intersects U, (cf. 4.1.8.). As in the proof of 4.2.3., we then can find a C^1 curve c in $S^1 \times I$ which is transverse to $\partial(S^1 \times I)$ and to F. As F is transversely orientable there exists a C^1 vector field Z on $S^1 \times I$ which is transverse to F and to $\partial(S^1 \times I)$, and which has c as an integral curve. The foliation ξ' defined by Z does not have a circle leaf. Hence the lemma follows from 6.3.1. □

A foliation $F \in F^1(\xi)$ with $C(F) = B_o$ can be defined by a projectable vector field $X \in X^1(\xi)$ which is transverse to B_1. The first-return map f of F, as introduced in 6.2.3., is here only a C^1 embedding, with the point 0 as an isolated fixed point. Moreover, the germ of f at 0 is a generator of the holonomy group of B_o and f is below or above the diagonal according as X points inwards or outwards on B_1.

6.3.3.- *Lemma*.- Let X and X' be projectable vector fields defining the foliations F and F' of $F^1(S^1 \times I)$ with B_o as unique circle leaf. Then F and F' are isotopic rel B_1 if and only if X and X' both point inwards or both point outwards on B_1.

Proof : By 6.3.2., we need only consider the case where $F, F' \in F^1(\xi)$. If F and F' are isotopic then their first-return maps are both above or both below the diagonal. Thus X and X' both point inwards or both point outwards on B_1.

For the proof of the converse let us consider the case where X and X' point inwards, the other case being similar. Let

$$F : S^1 \times (0,1] \to S^1 \times (0,1]$$

be the map which assigns to the point $z = \varphi_t(s)$, $s \in B_1$, $t \geq 0$, the point $F(z) = \varphi'_t(s)$. Then F is a homeomorphism which is the identity on B_1. Since F preserves each fibre of ξ it can be extended to B_o by the identity thus giving a homeomorphism which is isotopic to the identity and which takes $\mathcal{F}$ to $\mathcal{F}'$. □

6.3.4.- _Remark and Definition._ Now it is easily seen that each $\mathcal{F} \in \mathcal{F}^1(S^1 \times I)$ with $C(\mathcal{F}) = B_o$ is homeomorphic (but not necessarily isotopic) to half of a Reeb component as introduced in 3.3.1. We therefore call any such foliation _a half Reeb component_.

Next we can give a simple classification of all foliations of $\mathcal{F}^1(S^1 \times I)$ which are transverse to the boundary. But before we state the result in 6.3.7, we give a criterion for the differentiability of a foliation which is obtained by gluing together C^1 foliations along boundary components (cf. 3.5).

We recall that the notion of infinitesimal holonomy has been introduced in 5.3., exercise. The infinitesimal holonomy of order 1 is called _linear holonomy_ ; it is denoted by Dhol.

Let $(\Sigma_i,\mathcal{F}_i)$ be foliated surfaces of class C^1 and let L_i be a circle leaf of $\mathcal{F}_i$ in the boundary of Σ_i, $i = 1,2$. We denote by $(\Sigma,\mathcal{F})$ the foliation obtained by gluing $(\Sigma_1,\mathcal{F}_1)$ to $(\Sigma_2,\mathcal{F}_2)$ by means of a diffeomorphism $\varphi : L_1 \to L_2$.

6.3.5.- *Lemma*.- <u>The foliation</u> F <u>is an element of</u> $F^1(\Sigma)$ <u>if and only if</u> $F_i \in F^1(\Sigma_i)$, $i = 1,2$ <u>and</u> $Dhol_{L_1} = Dhol_{L_2}$.

Proof : Let L be the leaf of F which is obtained by identifying L_1 and L_2. We cover L by two distinguished squares U_o, U_1, as indicated in 3.2.5., figure 13. For both U_o and U_1 the transverse coordinate can be chosen in such a way that the only non-trivial coordinate change (if any) is given by a generator of the holonomy of L. This shows that $F \in F^1(\Sigma)$ if and only if the holonomy of L is of class C^1, (i.e. has its values in $G^1(\mathbb{R},0)$).

On the other hand, after a suitable choice of indices, we have

$$hol^+_L = hol_{L_1} \quad \text{and} \quad hol^-_L = hol_{L_2} \quad ;$$

see 3.2.8., vii). Thus hol_L is of class C^1 if and only if $Dhol_{L_1} = Dhol_{L_2}$. □

6.3.6.- *Remark*.- Of course, the same argument as in 6.3.5. holds also when two boundary leaves of the same foliation are identified.

Now coming back to the annulus, the following proposition, together with 6.3.1., gives a classification of foliations which are transverse to the boundary.

6.3.7.- *Proposition*.- <u>Let</u> $F \in F^1(S^1 \times I)$ <u>be a foliation which is transverse to the boundary and such that</u> $C(F) \neq \emptyset$.

i) <u>If</u> $C(F)$ <u>consists of just one leaf then</u> F <u>is obtained by gluing together two half Reeb components</u>.

ii) Suppose that $\mathcal{C}(\mathcal{F})$ consists of more than one leaf. Then there exists an annulus A in $S^1 \times I$ and a foliation $\mathcal{F}_1 \in \mathcal{F}^1(A)$ tangent to the boundary, such that $\mathcal{F}$ is obtained by attaching a half Reeb component along each boundary circle of A.

Proof : It is easily seen that for $j = 0,1$ there is $L_j \subset \mathcal{C}(\mathcal{F})$ such that L_j is contained in the closure of each leaf L of $\mathcal{F}$ which meets the boundary circle $B_j = S^1 \times \{j\}$. We cut $S^1 \times I$ along L_o and L_1.

i) If $L_o = L_1$ then we get two half Reeb components.

ii) If $L_o \neq L_1$ then we get three annuli, one of which, A, carries a foliation $\mathcal{F}_1$ which is tangent to the boundary. □

B) At this moment we stop our discussion of foliations on the annulus and come back to our initial problem of characterizing the structurally stable foliations on the annulus and on the torus. From now on, as in 6.2., Σ is either the annulus or the torus and ξ is the product fibration of Σ over S^1.

6.3.8.- *Definition*.- A circle leaf of a C^1 foliation $(\Sigma,\mathcal{F})$ is called hyperbolic if its linear holonomy is different from one. The set $\mathcal{C}(\mathcal{F})$ of circle leaves is hyperbolic if it is non-empty and consists of hyperbolic leaves only. (Note that for $\mathcal{F} \in \mathcal{F}^1(\xi)$ this definition coincides with 6.2.9.). In this case we also say that $\mathcal{F}$ is hyperbolic.

6.3.9.- *Lemma*.- Let $\mathcal{F} \in \mathcal{F}^1(\Sigma)$ be a foliation without Reeb components which is transverse to the boundary when $\partial\Sigma \neq \emptyset$. Then if $\mathcal{C}(\mathcal{F}) \neq \emptyset$ is non-hyperbolic, $\mathcal{F}$ is unstable.

Proof : If $C(F)$ is a single leaf L then, by 6.3.7., F is obtained by gluing together two half Reeb components along non-hyperbolic leaves. We then can "thicken L" by cutting Σ along L and filling in a product foliation by circles as indicated in figure 23.

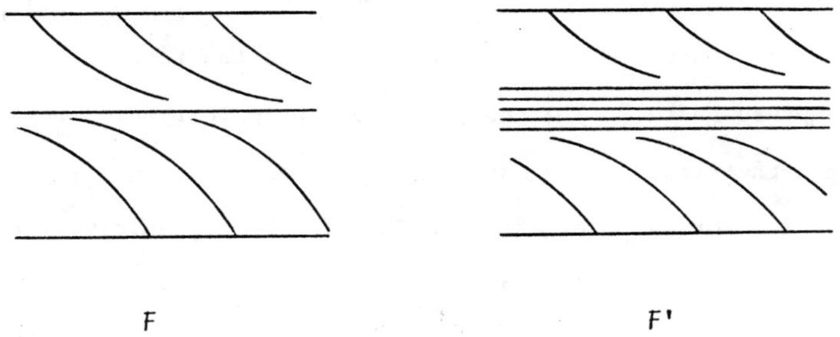

Figure 23

By 6.3.5., the foliation we obtain by this process is of class C^1 and has a tangent C^1 vector field, i.e. it is diffeomorphic to a foliation $F' \in F^1(\Sigma)$. Moreover, in this way we can construct foliations F' which are arbitrarily close to F. In other words, F cannot be stable.

In the other case, we use the fact that there is no Reeb component in F. Similar to 6.3.7. ii) we can find a foliation (A,F_1) in (Σ,F) which has a non-hyperbolic leaf and which is diffeomorphic to a foliation transverse to the fibration ξ. The last statement can be verified by showing that 4.2.15. holds also in the C^1 context.

Let $f_1 \in \mathrm{Diff}^1_+(I)$ be the first-return map of F_1. Proceeding as in section 6.2., we can approximate F_1 by a foliation F_2 whose first-return map f_2 has the properties that $Fix(f_2)$ and $Fix(f_1)$ are not conjugate and $Df_1(i) = Df_2(i)$ for $i = 0,1$; see 6.1.6. These properties of F_2 permit us to glue together (A, F_2) and $F|\Sigma - \mathring{A}$ to obtain a foliation of $F^1(\Sigma)$ which is close to F showing that F is unstable. □

6.3.10.- *Remark and definition*.- Let $F \in F^1(S^1 \times I)$ be a foliation which is transverse to the boundary and to the fibration ξ. Furthermore, suppose that $C(F)$ consists of a single leaf which is hyperbolic. Then the first-return map f of F has a unique fixed point x which is, moreover, hyperbolic. Thus the projectable vector field X defining F either points inwards or points outwards on both boundary circles ; cf. fig. 24.

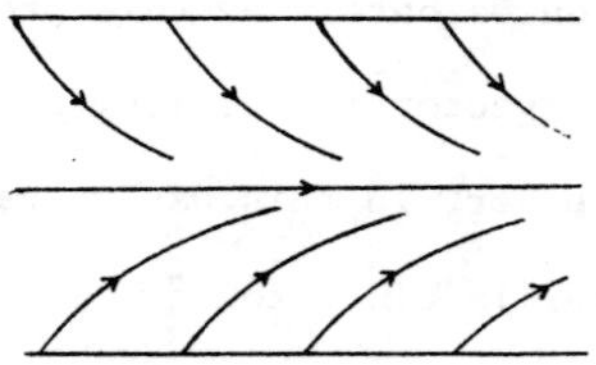

... X points inwards

Figure 24

In the first case we call x an attracting fixed point, in the second case an expanding fixed point of f. This is clearly equivalent to saying that $Df(x) < 1$ or $Df(x) > 1$, respectively.

The chief point in the proof of our main result 6.3.12. is provided by the following technical lemma.

6.3.11.- Lemma.- A foliation $\mathcal{F}$ as in 6.3.10. is structurally stable. Moreover, every foliation $\mathcal{F}'$ which is sufficiently close to $\mathcal{F}$ is conjugate to $\mathcal{F}$ by a homeomorphism which is the identity on the boundary.

Proof : We may assume that the projectable tangent vector field X points inwards on the boundary. Then the first-return map f is a C^1 diffeomorphism of I onto its image. Proceeding as in section 6.2., we can see that there is a neighbourhood W of $\mathcal{F}$ in $\mathcal{F}^1(S^1 \times I)$ such that the first-return map of each $\mathcal{F}' \in W$ has a unique fixed point which is moreover attracting. It follows that $\mathcal{F}'$ has a unique circle leaf. Hence, by 6.3.7., i), $\mathcal{F}'$ is obtained by gluing together two half Reeb components. Now a diffeomorphism F between $\mathcal{F}$ and $\mathcal{F}'$ which is the identity on the boundary can be constructed as in the proof of 6.3.3. □

Now we can give the desired characterization of the structurally stable foliations of $\Sigma = S^1 \times I$ or T^2.

6.3.12.- Theorem.- A foliation $\mathcal{F} \in \mathcal{F}^1(\Sigma)$ is structurally stable if and only if it is hyperbolic, and transverse to the boundary when $\Sigma = S^1 \times I$.

Proof : For simplicity, we consider only the case $\Sigma = T^2$. The proof of the case $\Sigma = S^1 \times I$ is left to the reader.

If $\mathcal{F} \in \mathcal{F}^1(T^2)$ is hyperbolic then $C(\mathcal{F})$ consists of a finite number of leaves and the restriction of $\mathcal{F}$ to the closure of each component R_j of $T^2 - C(\mathcal{F})$ is either a suspension or a Reeb component. In both cases there exists a closed transversal

θ_j in $\overset{o}{R}_j$. These transversals decompose T^2 into a finite number of annuli A_j where the restriction F_j of F to A_j satisfies the hypothesis of 6.3.11., thus is stable. As any foliation of $F^1(A_j)$ can be extended to a foliation of $F^1(T^2)$, we conclude that there is a neighbourhood W of F in $F^1(T^2)$ such that for each $F' \in W$ the restriction F'_j of F' to A_j is conjugate to F_j by a homeomorphism F_j which is the identity on ∂A_j ; cf 6.3.11. Thus the F_j fit together to give a homeomorphism between F and F'. Hence F is stable.

Conversely, if F is stable then by 6.2.8., $C(F) \neq \emptyset$ and $C(F) \neq T^2$. Moreover, by 4.2.14., ii), there is only a finite number of Reeb components in F and by cutting along a closed transversal in each of these Reeb components (if there is any) we can achieve a situation as in 6.3.9. It follows that F is hyperbolic. □

There is an analogue of the results 6.1.4., 6.1.5. and 6.1.10. for foliations on the annulus and foliations on the torus which should not be passed over. We leave its proof to the reader.

6.3.13.- *Theorem*.- For Σ the annulus or the torus the subset of structurally stable foliations is open and dense in $F^1(\Sigma)$.

To end this paragraph we treat briefly structural stability for foliations on the Möbius band and the Klein bottle, as well as structural stability for non-orientable foliations.

6.3.14.- *Exercises*. i) (Structural stability for orientable foliations on the Möbius band $\Sigma = S^1 \times_{\mathbb{Z}_2} I$ and the Klein bottle $\Sigma = S^1 \times_{\mathbb{Z}_2} S^1$).

a) Define the space $F^1(\Sigma)$ and the notion of structural stability for elements of $F^1(\Sigma)$, as in 6.2.1.

b) Extend the concept of hyperbolicity to orientation reversing diffeomorphisms of I and S^1 .

c) Using exercise 6.2.12, characterize the structurally stable foliations of $F^1(\Sigma)$.

d) Show that the subspace of structurally stable foliations is open and dense in $F^1(\Sigma)$.

ii) (Structural stability for non-orientable foliations on compact surfaces). - Let Σ be one of $S^1 \times K$, $S^1 \times_{\mathbb{Z}_2} K$, $K = I$ or S^1 . A <u>line field</u> D on Σ assigns to each $x \in \Sigma$ a 1-dimensional subspace of the tangent space $T_x\Sigma$ at the point x . Say that a line field D is of <u>class</u> C^r if, for each $x \in \Sigma$, there is a neighbourhood U of x and a C^r vector field X_U on U which spans D , i.e. D(y) is spanned by $X_U(y)$ for each $y \in U$.

a) Show that there ia an injective map from the set of C^{r+1} foliations on Σ to the set of C^r line fields of Σ .

b) Using a), define the topology of C^1 uniform convergence on the set of all C^1 foliations on Σ which admit a C^1 tangent line field. Do this in such a way that $F^1(\Sigma)$, with its original topology, becomes a subspace (in fact a connected component).

c) Define structural stability for non-orientable foliations and give a characterization of the structurally stable foliations.

d) Using the concept of the tangent orientation covering, try to reduce structural stability for non-orientable foliations to that for orientable foliations.

iii) If Σ is as above, show that the set of all fibrations of Σ over S^1 is not closed in $F^1(\Sigma)$.

CHAPTER II

FUNDAMENTALS ON FOLIATIONS

1. *Foliated bundles*.

In this chapter the central subject of this book is presented in full generality. Before we give (in 2.1) the definition we study an intermediate class of objects, the so-called foliated bundles. This is for three reasons:

Foliated bundles are fibre bundles with a foliation transverse to the fibres (or with a flat connection, in the language of differential geometry) and so may help the reader - at least the reader familiar with fibre bundles - to better visualize the subsequent "general" definition of foliations (§ 2).

They are on the other hand an important class of foliations and the method of their construction by "suspension of group actions", as developed in section 1.2, is one of the main construction principles in foliation theory.

Last but not least, the study of foliated bundles may be considered as a starting point of one of the main branches of foliation theory, namely that of classifying spaces. (This topic will only be touched in this book.) The names of Ehresmann, Haefliger, Thurston (and many others) must be mentioned here.

As a further step to motivate the general concept we enlarge, in 1.4, the class of foliated bundles in a natural way. It turns out that among the "foliations obtained by an equivariant submersion" one can find the rich class of foliations with a transverse structure (see section 3.2) and, as prime example of a foliation, the Reeb component on $S^1 \times D^{m-1}$.

We need some notation and terminology. These will be given in the form of a brief review of some basic material concerning fibre bundles. For background on fibre bundles and more detailed information we refer the reader to the books of Hirzebruch [Hir], Husemoller [Hu] or Steenrod [St], for example.

1.1. *Preparatory material on fibre bundles.*

In this paragraph B, F and M denote connected manifolds. We allow F and M to have a boundary.

1.1.1. - *Definition.* - Let $p : M \to B$ be a continuous map. The triple $\xi = (M,p,B)$ is called a (locally trivial) fibre bundle, with total space M, base B and fibre F if there is an open covering $\{U_i\}$ of B and homeomorphisms

$$\phi_i : p^{-1}(U_i) \longrightarrow U_i \times F$$

such that the following diagram is commutative (local triviality) :

$$\begin{array}{ccc} p^{-1}(U_i) & \xrightarrow{\ \phi_i\ } & U_i \times F \\ & {\scriptstyle p|p^{-1}(U_i)} \searrow \quad \swarrow {\scriptstyle pr_1} & \\ & U_i & \end{array}$$

where pr_1 is the projection onto the first factor.

The simplest example of a fibre bundle evidently is $(B \times F, pr_B, B)$ which is called the product bundle.

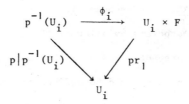

A pair (U_i, ϕ_i) is referred to as a <u>local trivialization</u> of ξ. For each $b \in B$, the set $F_b = p^{-1}(b)$ is called the <u>fibre over</u> b. It is homeomorphic to F by the restriction ϕ_{ib} of ϕ_i to F_b, where $b \in U_i$.

The family $A = \{(U_i, \phi_i)\}$ of local trivializations is called a <u>fibre bundle atlas</u> of ξ. Clearly, any fibre bundle admits many fibre bundle atlases. But any two of them are equivalent in the sense that their union is again a fibre bundle atlas.

1.1.2. - <u>*Definition*</u>. - Let $\xi = (M,p,B)$ and $\xi' = (M',p',B')$ be fibre bundles with fibre F. A continuous map $f : M \to M'$ is a <u>fibre bundle map</u> if

(1) f preserves the fibres, i.e. f induces a map $\bar{f} : B \to B'$ such that $p' \circ f = \bar{f} \circ p$,

(2) for every $b \in B$, the restriction f_b of f to F_b carries F_b homeomorphically onto $F_{\bar{f}(b)}$.

The fibre bundle map f is a <u>fibre bundle isomorphism</u> (<u>over</u> B) if $B' = B$, $\bar{f} = \mathrm{id}$. (In particular, f is a homeomorphism and f^{-1} is also a fibre bundle isomorphism).

1.1.3. - <u>*Remark*</u>. - It is possible to define a somewhat broader notion of fibre bundle morphisms by allowing the fibre F to vary or by admitting maps which are not necessarily homeomorphisms on the fibres. Also fibre bundle isomorphisms could be defined for bundles over different bases. The definitions chosen seem, however, to be the ones most suitable for our purposes.

Only in a very special situation shall we consider a more general class of bundle morphisms. This, however, will not be until paragraph 2.3. and is particularly emphazised there.

1.1.4. - Let (U_i,ϕ_i) and (U_j,ϕ_j) be two local trivializations of ξ. If $U_i \cap U_j \neq \emptyset$ we get the following commutative diagram

$$\begin{array}{ccccc} (U_i \cap U_j) \times F & \xleftarrow{\ \phi_j\ } & p^{-1}(U_i \cap U_j) & \xrightarrow{\ \phi_i\ } & (U_i \cap U_j) \times F \\ & {\scriptstyle pr_1} \searrow & \downarrow {\scriptstyle p} & \swarrow {\scriptstyle pr_1} & \\ & & U_i \cap U_j & & \end{array}$$

and the map

$$\begin{aligned} \phi_{ij} : U_i \cap U_j &\longrightarrow \text{Homeo}(F) \\ b &\longmapsto \phi_{ib} \circ \phi_{jb}^{-1} \end{aligned}$$

satisfies the following two conditions

(1) $\phi_{ii}(b) = id_F$, for every i and any $b \in U_i$

(2) If $U_i \cap U_j \cap U_k \neq \emptyset$ then

$$\phi_{ik}(b) = \phi_{ij}(b) \circ \phi_{jk}(b), \quad \text{for any } b \in U_i \cap U_j \cap U_k.$$

The map ϕ_{ij} is called the <u>coordinate transformation</u> of A over $U_i \cap U_j$ and the pair $C = (\{U_i\},\{\phi_{ij}\})$ is called the <u>cocycle corresponding to the fibre bundle atlas</u> A.

For many fibre bundles there is an additional structure which may be described by the following data :

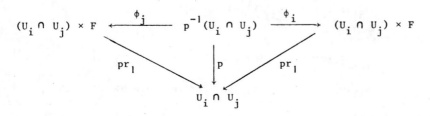

(1) A fibre bundle atlas $\mathcal{A} = \{(U_i,\phi_i)\}$,

(2) a topological group G ,

(3) an effective continuous action $A : G \times F \to F$,

(4) for every pair (i,j) such that $U_i \cap U_j \neq \emptyset$ a continuous map

$$g_{ij} : U_i \cap U_j \to G$$

such that

$$\phi_{ij}(b)(y) = A(g_{ij}(b),y) \text{ for all } y \in F.$$

Note that

$$g_{ii}(b) = id_G \text{ for every } i \text{ and all } b \in U_i$$

and

$$g_{ik}(b) = g_{ij}(b) \circ g_{jk}(b) \text{ for any } b \in U_i \cap U_j \cap U_k ,$$

by the effectivity of A .

The pair $C = (\{U_i\},\{g_{ij}\})$ is called a (continuous) cocycle on B with values in G .

1.1.5. - Definition.

i) The fibre bundle atlas $\mathcal{A}$ is called a G-atlas if its corresponding cocycle is continuous and has values in G, in the sense of (1) - (4).

ii) Two G-atlases $\mathcal{A}$ and $\mathcal{A}'$ are equivalent if their union is again a G-atlas.

iii) A fibre bundle together with an equivalence class of G-atlases (maximal G-atlas or a G-structure) is called a fibre bundle with structure group G or briefly a G-bundle.

A G-structure on a fibre bundle ξ is defined once there is given a G-atlas of ξ. For example, the product bundle $(B \times F,pr_B,B)$ provided with the trivial G-atlas is a G-bundle, for every topological group G acting effectively on F.

As will be shown later, even product bundles can be endowed with different G-structures, cf. 1.2.13. and 1.3.11. i).

We now fix once and for all an effective continuous action of the topological group G on F and identify G with its image in Homeo(F) by means of this action.

As for G-atlases there is also a notion of equivalence for cocycles on B with values in G.

1.1.6. - *Definition*. - The cocycles $C = (\{U_i\},\{g_{ij}\})$ and $C' = (\{U'_\kappa\},\{g'_{\kappa\lambda}\})$ on B with values in G are equivalent (or cohomologous) if for every pair (i,κ) such that $U_i \cap U'_\kappa \neq \emptyset$ there exists a continuous map

$$\bar{g}_{\kappa i} : U_i \cap U'_\kappa \to G$$

such that

(1) $\bar{g}_{\kappa i}(b) = \bar{g}_{\kappa j}(b) g_{ji}(b)$, for $b \in U_i \cap U_j \cap U'_\kappa$,

(2) $\bar{g}_{\lambda i}(b) = g'_{\lambda\kappa}(b)\bar{g}_{\kappa i}(b)$, for $b \in U_i \cap U'_\kappa \cap U'_\lambda$.

Roughly speaking this definition means that the cocycles C and C' are equivalent if their union is again a cocycle on B with values in G.

The equivalence class $[C]$ of a cocycle C will be referred to as a cohomology class.

1.1.7.- *Lemma*. - Let $A = \{(U_i,\phi_i)\}$ and $A' = \{(U'_\kappa,\phi'_\kappa)\}$ be G-atlases of the G-bundle $\xi = (M,p,B)$. Then the cocycles corresponding to A and A' are equivalent.

Proof : The fact that A and A' are equivalent means that

$$\bar{g}_{\kappa i}(b) = \phi'_{\kappa b}\circ\phi_{ib}^{-1}, \qquad b \in U_i \cap U'_\kappa$$

coincides with the operation of an element of G and the map

$$\bar{g}_{\kappa i} : U_i \cap U'_\kappa \to G$$

so obtained is continuous. It is readily proved that $\{\bar{g}_{\kappa i}\}$ satisfies the conditions (1) and (2) for the cocycles associated to A and A'. □

We also need morphisms between G-bundles:

1.1.8. - *Definition.* - Let $\xi = (M,p,B)$ and $\xi' = (M',p',B')$ be G-bundles with fibre F. A fibre bundle map $(f,\bar{f})$ between ξ and ξ', considered as fibre bundles, is a G-<u>bundle map</u> between ξ and ξ' if there are G-atlases $A = \{(U_i,\phi_i)\}$ and $A' = \{(U'_\kappa,\phi'_\kappa)\}$ of ξ and ξ', respectively, such that for every (i,κ) with $U_i \cap \bar{f}^{-1}(U'_\kappa) \neq \emptyset$ and every $b \in U_i \cap \bar{f}^{-1}(U'_\kappa)$ the homeomorphism

$$\bar{g}_{\kappa i}(b) = \phi'_{\kappa b'} \circ f_b \circ \phi_{ib}^{-1}, \quad b' = \bar{f}(b),$$

belongs to G and the map

$$\bar{g}_{\kappa i} : U_i \cap \bar{f}^{-1}(U'_\kappa) \to G$$

so defined is continuous. (It follows by 1.1.7. that checking this additional condition for $(f,\bar{f})$ may be done by use of any two G-atlases of ξ and ξ').

The notion of a G-<u>isomorphism</u> for G-bundles over the same base is now introduced in the obvious way. Note that if $f : \xi \to \xi'$ is a G-isomorphism then it easily follows by the continuity of $G \to G$, $g \mapsto g^{-1}$, that f^{-1} is also a G-isomorphism.

Note that a remark similar to 1.1.3. could be made here.

A G-bundle G-isomorphic to the product bundle is said to be trivial.

1.1.9. - Lemma. - The G-bundles $\xi = (M,p,B)$ and $\xi' = (M',p,B')$ with fibre F are G-isomorphic if and only if given any two G-atlases of ξ resp. ξ' their corresponding cocycles are equivalent.

Proof : Suppose $f : \xi \to \xi'$ is a G-isomorphism. This means that given any two G-atlases of ξ and ξ', respectively, there is a family of continuous maps $\{\bar{g}_{\kappa i}\}$ defined as in 1.1.8. Using these maps one easily verifies conditions (1) and (2) of 1.1.6.

Conversely, given G-atlases $A = \{(U_i,\phi_i)\}$ and $A' = \{(U'_\kappa,\phi'_\kappa)\}$ of ξ and ξ', respectively, and maps $\bar{g}_{\kappa i}$ satisfying 1.1.6. for the cocycles corresponding to A and A' we may define $f : M \to M'$ in the following way.

For $x \in p^{-1}(U_i \cap U'_\kappa)$ and $b = p(x)$ we set

$$f_{\kappa i}(x) = \phi'^{-1}_\kappa(b,\bar{g}_{\kappa i}(b)(\phi_{ib}(x))).$$

The maps $f_{\kappa i}$ are continuous and conditions (1) and (2) of 1.1.6. guarantee that two such maps coincide on the intersection of their domains. Thus there is defined a continuous map $f : M \to M'$ which is moreover bijective and thus a G-isomorphism. □

1.1.10. - Corollary. - Let ξ be a G-bundle with base B and fibre F. If $G = \{e\}$ then ξ is trivial.

1.1.11. - Fibre bundles defined by cocycles. - Given a cocycle $C = (\{U_i\},\{g_{ij}\})$ on B with values in G, a G-bundle $\xi_C = (M,p,B)$

with fibre F can be constructed as follows. We take the disjoint union $E = \coprod_i (U_i \times F)$ and consider the equivalence relation ρ :

$$(b,y)\ \rho\ (b',y') \quad \text{if} \quad b = b' \in U_i \cap U_j \quad \text{and} \quad y' = g_{ij}(b)(y).$$

Let $\pi : E \to M = E/\rho$ and $r : E \to B$ be the natural projections. Since r is compatible with ρ there is an induced map $p : M \to B$ which is a fibre bundle together with a G-atlas whose corresponding cocycle is C.

Clearly, if C is the trivial cocycle then ξ_C is the product bundle.

It follows by 1.1.9 that equivalent cocycles C and C' yield isomorphic G-bundles ξ_C and $\xi_{C'}$. Therefore, by the above construction, we have a map H from the set of cohomology classes of cocycles on B with values in G to the set of isomorphism classes of G-bundles with fibre F and base B and taking the cohomology class of the trivial cocycle to the isomorphism class of the product bundle. It follows from 1.1.9 that H is in fact a bijection.

Fibre bundles in the isomorphism class corresponding to the cohomology class $[C]$ under H are said to be <u>associated</u> to $[C]$.

If in particular $G = F$ and the action of G on itself is left translation then fibre bundles with structure group G and fibre G are called <u>principal bundles</u>.

1.1.12.- <u>The induced cocycle and induced bundle.</u>

Let $C = (\{U_i\},\{g_{ij}\})$ be a cocycle on B with values in G and let $\bar{f} : B' \to B$ be a continuous map. Then a cocycle $\bar{f}^*C = (\{U_i^*\},\{g_{ij}^*\})$ on B' with values in G is defined by $U_i^* = \bar{f}^{-1}(U_i)$ and $g_{ij}^* = g_{ij} \circ \bar{f}|(U_i^* \cap U_j^*)$. It is called the <u>cocycle induced</u> from C by the map $\bar{f}$.

If C and C' are equivalent cocycles on B then $\bar{f}^*C$

and $\bar{f}^*C'$ are equivalent cocycles on B'.

Given a G-bundle $\xi = (M,p,B)$ with fibre F associated to $[C]$, a G-bundle $\bar{f}^*\xi = (M',p',B')$ with fibre F associated to $[\bar{f}^*C]$ is constructed in the following way. Set

$$M' = \{(b',x) \in B' \times M \mid \bar{f}(b') = p(x)\}.$$

and define
$$\begin{aligned} p' : M' &\longrightarrow B' \\ (b',x) &\longrightarrow b'\ . \end{aligned}$$

Then $\bar{f}^*\xi$ is a fibre bundle. It is called the <u>fibre bundle induced from ξ by the map</u> $\bar{f}$. Any G-atlas $A = \{(U_i,\phi_i)\}$ of ξ may be used to define a G-atlas A^* of $\bar{f}^*\xi$. A local trivialization of A^* is given by

$$\begin{aligned} \phi_i^* : p'^{-1}(U_i^*) &\longrightarrow U_i^* \times F \\ (b',x) &\longrightarrow (b',\phi_{ib}(x)) \end{aligned}$$

where $b = \bar{f}(b')$. One easily verifies that A^* is in fact a G-atlas.

Note that the canonical map $f : M' \to M$ defined by $(b',x) = x$ makes the diagram

$$\begin{array}{ccc} M' & \xrightarrow{\ f\ } & M \\ {\scriptstyle p'}\big\downarrow & & \big\downarrow{\scriptstyle p} \\ B' & \xrightarrow{\ \bar{f}\ } & B \end{array}$$

a G-bundle map.

The induced bundle has the following "universal" property. Given a G-bundle $\eta = (\hat{M},\hat{p},B')$ and a G-bundle map $(\hat{f},\bar{f})$ between η and ξ there is a unique G-isomorphism $h : \eta \to \xi$ making a commutative diagram

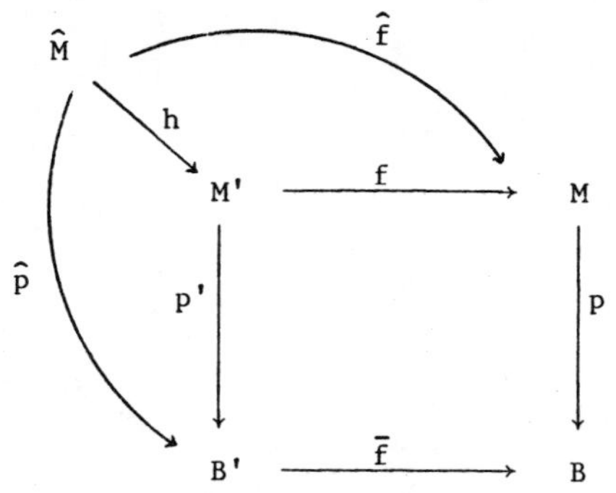

Remark.- The fact that h is a G-isomorphism is due to our special choice of the class of G-morphisms. The universal property of the induced bundle in fact holds for a broader class of fibre bundle morphisms. The map h is then however in general no longer an isomorphism.

As the construction of the induced bundle describes a certain change of the base of a fibre bundle there is also a construction which describes a certain change of the structure group.

1.1.13. - *Reduction of the structure group.*

Let $\xi = (M,p,B)$ be a G-bundle and let Γ be a subgroup of G. If there exists a G-atlas $A = \{(U_i,\phi_i)\}$ of ξ such that the corresponding cocycle $C = (\{U_i\},\{g_{ij}\})$ has values in Γ then ξ is also a Γ-bundle and we say that the structure group of ξ can be reduced to the subgroup Γ.

In particular, by 1.1.10, a G-Bundle is trivial if and only if its structure group can be reduced to $\{e\}$.

Conversely, if G is contained in G' then ξ may be considered as a G'-bundle. If two G-bundles are isomorphic then they are of course isomorphic as G'-bundles.

Let Γ_1 and Γ_2 be subgroups of G and let ξ_i be a Γ_i-bundle with base B and fibre F, $i = 1,2$. We say that ξ_1 and ξ_2 are G-isomorphic if they are isomorphic as G-bundles.

We shall be interested in the case where the subgroup Γ of G is totally disconnected, that is the connected components of Γ are the points of Γ.

As an easy consequence of the definition we have :

1.1.14. - Lemma. - The structure group of the G-bundle ξ can be reduced to a totally disconnected subgroup Γ if and only if there exists a G-atlas of ξ whose corresponding cocycle $(\{U_i\},\{g_{ij}\})$ is locally constant (i.e., each g_{ij} is locally constant) with values in Γ.

Proof : Clearly, we only have to show the "if" statement ; the "only if" statement follows directly by the definition and the fact that the g_{ij} are continuous.

Given a locally constant cocycle $(\mathcal{U},\{g_{ij}\})$ we may first take a countable subcover $\{U_\kappa\}$ of $\mathcal{U}$. The set of connected components of the intersections $U_\kappa \cap U_\lambda$ is countable. As the g_{ij} are locally constant it follows that the set

$$\{g \in G \mid g = g_{\kappa\lambda}(b) \text{ for some } g_{\kappa\lambda} : U_\kappa \cap U_\lambda \to G \text{ and } b \in U_\kappa \cap U_\lambda\}$$

is countable and thus generates a countable subgroup of G. The lemma now follows by the observation that every countable topological group is totally disconnected. □

1.2. Suspensions of group actions.

We have already defined (see I; 3.2.) what the suspension of a homeomorphism of S^1 is. The same construction can be made using homeomorphisms of arbitrary manifolds. It is nothing else than the mapping torus. We are now going to generalize this construction.

The meaning of B and F is unchanged during this section. The topological group G is now $\mathrm{Homeo}(F)^{\delta}$ the group $\mathrm{Homeo}(F)$ of self-homeomorphisms of F endowed with the discrete topology. By

$$q : (\tilde{B},\tilde{b}_o) \rightarrow (B,b_o)$$

we denote the universal covering of B with respect to base points b_o and $\tilde{b}_o$.

1.2.1. - Suspension. - Let

$$H : \pi_1(B,b_o) \rightarrow \mathrm{Homeo}(F)$$

be a representation. If we identify $\pi_1(B,b_o)$ with the group of covering translations of the universal covering q then we get an action of $\pi_1(B,b_o)$ on $\tilde{B} \times F$ which is defined as follows:

$$\begin{aligned} A : \pi_1(B,b_o) \times (\tilde{B} \times F) &\rightarrow (\tilde{B} \times F) \\ (\gamma,(\tilde{b},y)) &\mapsto (\gamma(\tilde{b}),H(\gamma)(y)) \end{aligned}$$

This action yields a commutative diagram

$$\begin{array}{ccc} \tilde{B} \times F & \xrightarrow{\ pr\ } & \tilde{B} \\ \pi \big\downarrow & & \big\downarrow q \\ M = (\tilde{B} \times F)/A & \xrightarrow{\ p\ } & B \end{array} \qquad (*)$$

where pr is the canonical projection, π is the quotient map by A , M is the quotient space, and p is uniquely induced by pr .

This construction is called suspension (with fibre F) of the representation H. We shall also say that (M,p,B) is the suspension of the representation H .

Suspensions are of great importance in the theory of foliations. The rest of this paragraph is devoted to the study of their geometric properties.

1.2.2. - *Proposition*. - <u>In the diagram</u> (*)

i) <u>the map</u> $\pi : \tilde{B} \times F \to M$ <u>is a covering map</u> ;

ii) <u>if</u> $\Gamma = \text{im } H$ <u>is endowed with the induced topology then</u> $\xi_H = (M,p,B)$ <u>is a fibre bundle with fibre</u> F <u>and discrete structure group</u> Γ ;

iii) <u>the induced</u> Γ-<u>bundle</u> $q^*\xi_H$ <u>is trivial</u>.

Proof : To see that π is a covering map it suffices to notice that the action A of $\pi_1 B$ on $\tilde{B} \times F$ is free and properly discontinuous, (cf. Wolf [Wo]).

For the proof of ii) we choose

(1) a covering $\{U_i\}$ of B by contractible open sets U_i,

(2) for each i a point $b_i \in U_i$ and a path c_i from b_o to b_i.

Now for each U_i we want to find a local trivialization $\phi_i : p^{-1}(U_i) \to U_i \times F$. For that let $\tilde{c}_i$ be the lift of c_i under q starting in $\tilde{b}_o$ and let $\tilde{U}_i$ be the component of $q^{-1}(U_i)$ which contains the endpoint of $\tilde{c}_i$. Since every non-trivial element of $\pi_1(B,b_o)$ permutes all components of $pr^{-1}(q^{-1}(U_i))$ it follows that

$$\pi_i = \pi|(\tilde{U}_i \times F) : \tilde{U}_i \times F \to p^{-1}(U_i)$$

is a homeomorphism. Then

$$\phi_i = (q|\tilde{U}_i \times id_F) \circ \pi_i^{-1} : p^{-1}(U_i) \to U_i \times F$$

is a local trivialization of ξ_H over U_i, by the commutativity of (*).

Let $U_i \cap U_j \neq \emptyset$. To every $b \in U_i \cap U_j$ we introduce paths w_{ib} in U_i from b_i to b and w_{jb} from b_j to b, see fig. 1.

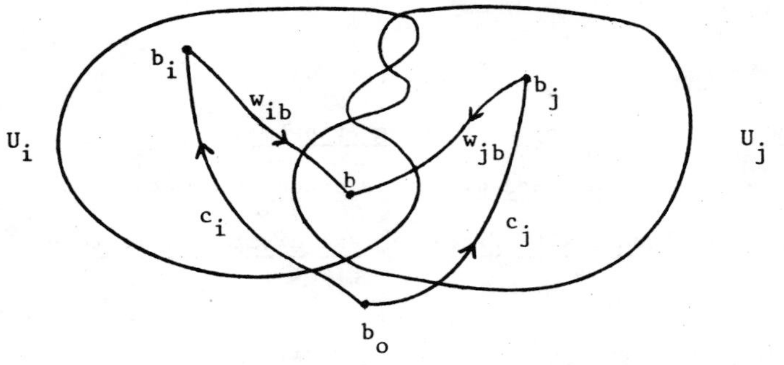

Figure 1

Clearly, the element $\gamma_{ij} = [c_i * w_{ib} * w_{jb}^{-1} * c_j^{-1}] \in \pi_1(B,b_o)$ does not depend on the choice of w_{ib} and w_{jb} and so the map

$$\begin{aligned} g_{ij} : U_i \cap U_j &\to \Gamma \\ b &\mapsto H(\gamma_{ij}) \end{aligned}$$

is locally constant (and so in particular continuous). By construction, $(\{U_i\},\{g_{ij}\})$ is a cocycle on B with values in Γ. This proves ii).

Statement iii) is an immediate consequence of the universal property of the induced bundle. □

1.2.3. - Leaf topology.

For the study of the geometric properties of suspensions we introduce a new topology for the total space of $\xi_H = (M,p,B)$. We denote by F^δ the set F supplied with the discrete topology. Then the action A of $\pi_1 B$ on $\tilde{B} \times F^\delta$ remains continuous and the map $\pi : \tilde{B} \times F^\delta \to M$ induces on M a new topology which is finer than its manifold topology. Denote by M^δ the set M supplied with this topology. The topology on $\tilde{B} \times F^\delta$ and the topology on M^δ are called the <u>leaf topologies</u>, (the word <u>fine topology</u> can also be found in the literature). Note that the projections $\pi : \tilde{B} \times F^\delta \to M^\delta$ and $p : M^\delta \to B$ are still continuous, the first by definition of M^δ.

We immediately have :

1.2.4. - Lemma. - The suspension diagram

$$\begin{array}{ccc} \tilde{B} \times F^\delta & \xrightarrow{\ pr\ } & \tilde{B} \\ \pi \downarrow & & \downarrow q \\ M^\delta & \xrightarrow[\ p\ ]{} & B \end{array}$$

is a commutative diagram of covering maps.

1.2.5. - Remarks and notation. - The topological space M^δ is non-connected unless the fibre reduces to a point. A connected component of M^δ is called a <u>leaf</u> of ξ_H. Each point $x = \pi(\tilde{b},y) \in M$ belongs to exactly one leaf which is denoted by L_x and equals $\pi(\tilde{B} \times \{y\})$. Clearly, there are uncountably many leaves if F is not a point.

The leaves are injectively immersed submanifolds of M but in general not embedded (see examples below). They are transverse to the fibres of ξ_H.

Note that every suspension with fibre F is a $\mathrm{Homeo}(F)^\delta$-bundle.

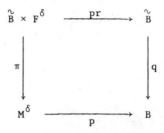

We next want to see to what extent the suspension ξ_H depends on the representation H. For this let ξ_H and $\xi_{H'}$ be the suspensions of two representations H and H' , with the same fibre F but base spaces B, B', respectively. (The data used for ξ_H are those of 1.2.1, and similarly for $\xi_{H'}$ indicated by '). Let

$$\tilde{f} = (f_o, g) : \tilde{B} \times F \to \tilde{B}' \times F$$

be a continuous product map such that

(1) $g : F \to F$ is a homeomorphism,

(2) $\tilde{f}$ is compatible with the actions A and A'.

Then $\tilde{f}$ induces maps f and $\bar{f}$ making the following diagram commutative

$$\begin{array}{ccc}
\tilde{B} \times F & \xrightarrow{\tilde{f}} & \tilde{B}' \times F \\
\downarrow{\scriptstyle \pi} & & \downarrow{\scriptstyle \pi'} \\
M & \xrightarrow{f} & M' \\
\downarrow{\scriptstyle p} & & \downarrow{\scriptstyle p'} \\
B & \xrightarrow{\bar{f}} & B'
\end{array}$$

and such that $(\tilde{f}, f)$ and $(f, \bar{f})$ are homomorphisms of coverings with respect to the leaf topologies.

1.2.6. - *Definition*.- The triple $(\tilde{f}, f, \bar{f})$ is called a map of suspensions (or briefly, an S-map).

The notion of an S-isomorphism is then introduced in the obvious way.

1.2.7. - *Lemma*. - Let $(\tilde{f}, f, \bar{f})$ be an S-map (resp. S-isomorphism, if $B = B'$ and $\bar{f} = id$). Then $(f, \bar{f}) : \xi_H \to \xi_{H'}$ is a $\mathrm{Homeo}(F)^{\delta}$-bundle map (resp. $\mathrm{Homeo}(F)^{\delta}$-isomorphism).

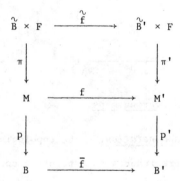

Proof : Clearly, $(f,\bar{f})$ is a fibre bundle map. Moreover, using Homeo(F)-atlases $A = \{(U_i,\phi_i)\}$ and $A' = \{(U'_\kappa,\phi'_\kappa)\}$ of ξ_H and $\xi_{H'}$, as introduced in the proof of 1.2.2., the maps

$$\bar{g}_{\kappa i} : U_i \cap \bar{f}^{-1}(U'_\kappa) \to \mathrm{Homeo}(F)^\delta$$

$$\bar{g}_{\kappa i}(b) = \phi'_{\kappa b'} \circ f_b \circ \phi_{ib}^{-1} , \quad b' = \bar{f}(b),$$

are locally constant and therefore continuous. This shows that $(f,\bar{f})$ is a $\mathrm{Homeo}(F)^\delta$-bundle map, or a $\mathrm{Homeo}(F)^\delta$-isomorphism, if $B = B'$ and $\bar{f} = id$. □

We shall see (1.3.1) that a converse of 1.2.7. is also true.

1.2.8. - Lemma. - <u>There is a one-one correspondence between</u>

(1) <u>the set of S-maps from</u> ξ_H <u>to</u> $\xi_{H'}$ <u>and</u>

(2) <u>the set of continuous maps</u> $\bar{f} : (B,b_o) \to (B',b'_o)$ <u>such that there exists</u> $g \in \mathrm{Homeo}(F)$ <u>with</u>

$$H(\gamma) = g^{-1}H'(\bar{f}_\#(\gamma))g, \quad \text{for any } \gamma \in \pi_1(B,b_o).$$

(We then say that H and H' are <u>conjugate</u>).

Proof : The set of S-maps between ξ_H and $\xi_{H'}$ is by definition in one-one correspondence with the set of continuous maps

$$\tilde{f} = (f_o,g) : \tilde{B} \times F \to \tilde{B}' \times F$$

which are compatible with the actions A and A' on $\tilde{B} \times F$ and $\tilde{B}' \times F$, respectively. The compatibility condition for $\tilde{f}$ means that, for any $\gamma \in \pi_1(B,b_o)$, we have a commutative diagram

$$\begin{array}{ccc} \tilde{B} \times F & \xrightarrow{\tilde{f}} & \tilde{B}' \times F \\ \downarrow{\scriptstyle (\gamma, H(\gamma))} & & \downarrow{\scriptstyle (\bar{f}_\#(\gamma), H'(\bar{f}_\#(\gamma)))} \\ \tilde{B} \times F & \xrightarrow{\tilde{f}} & \tilde{B}' \times F \end{array}$$

Restricted to the factor F this yields $gH(\gamma) = H'(\bar{f}_\#(\gamma))g$, as required.

Conversely, given $\bar{f} : (B,b_o) \to (B',b_o')$ we lift it to the universal coverings $f_o : \tilde{B} \to \tilde{B}'$ and set $\tilde{f} = (f_o, g)$ where g is given by (2). Then $\tilde{f}$ is compatible with the actions A and A' and thus is an S-map. □

As an immediate consequence of the last lemma we get the following result.

1.2.9. - *Proposition*. - Let $H, H' : \pi_1(B,b_o) \to \mathrm{Homeo}(F)$ be two representations. Then their suspensions are S-isomorphic if and only if H and H' are conjugate.

Remark.- A change of the basepoint $b_o \in B$ has the consequence that the representation H has to be replaced by a conjugate one.

Before we give any examples of suspensions we introduce another important concept relating the leaves of a suspension ξ_H with the representation H.

First a lemma.

1.2.10. - *Lemma* (unexplained notation as in 1.2.1).

Let ξ_H be a suspension and let L_x be the leaf passing _through_ $x = \pi(\tilde{b}_o, y)$. _Then_ $\pi_1(L_x, x)$ _is isomorphic to the isotropy group_ $\{\gamma \in \pi_1(B, b_o) \mid H(\gamma)(y) = y\}$ of $\pi_1(B, b_o)$ _in the point_ $y \in F$.

Proof : The lemma follows from the commutative triangle of covering maps

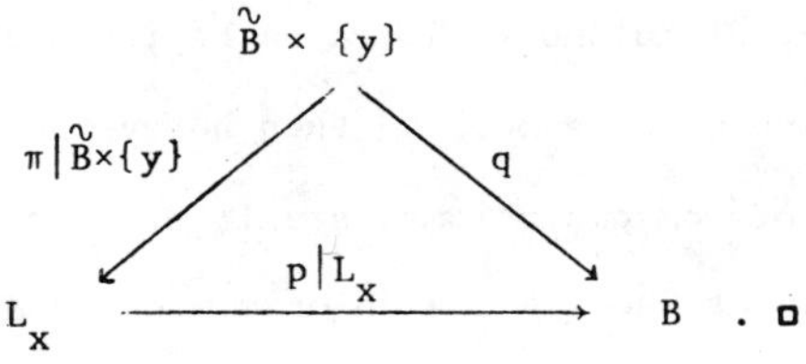

By the preceding lemma, we get the commutative diagram

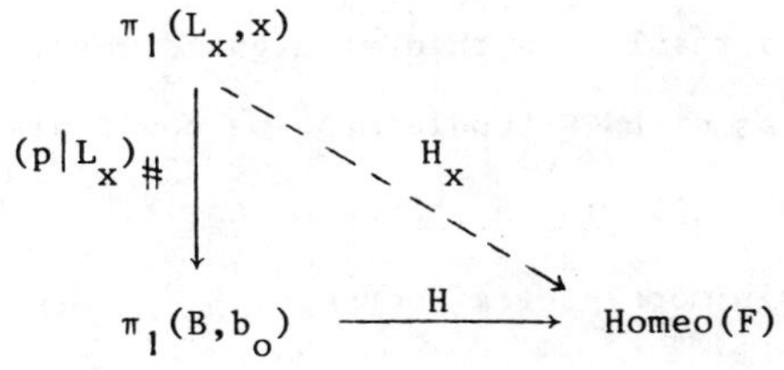

with $\operatorname{im} H_x = \Gamma_y$, where $\Gamma = \operatorname{im} H$ and $\Gamma_y = \{g \in \Gamma | g(y) = y\}$ the isotropy group of Γ in $y \in F$.

1.2.11. - Definition. - The representation

$$H_x : \pi_1(L_x, x) \to \text{Homeo}(F)$$

with image Γ_y is called the _holonomy_ representation _of the leaf_ L_x (in the point x). The group Γ_y is the _holonomy group of the leaf_ L_x (in the point x).

A change of the base point has the effect that H_x has to be replaced by a conjugate representation. The holonomy representation is of importance since it describes the behaviour of the leaves near L_x, by a theorem of Haefliger [Ha]. We shall describe this partially and provisionally in form of an exercise (see 1.2.14., iii)).

The concept of holonomy will also be introduced for leaves in arbitrary foliations. The group G then however has to be replaced by a group of germs of homeomorphisms, see I; 3.2., and III; 2.1, also for a discussion of Haefliger's theorem mentioned above.

1.2.12. - *Remarks*. - i) In all of this section $\mathrm{Homeo}(F)^\delta$ could be replaced by any totally disconnected subgroup G of $\mathrm{Homeo}(F)$, where $\mathrm{Homeo}(F)$ is endowed with an arbitrary topology. The modification of definitions and results is then straightforward.

ii) As with G-bundle maps we could also broaden the notion of an S-map by allowing a change of the fibre. We shall however not make use of this more general concept.

1.2.13. - *Examples*. i) We have seen in chapter I that there can occur very different situations with the suspension of just one homeomorphism of S^1. Recall for example that

(1) the leaves are homeomorphic to either S^1 or $\mathbb{R}$ and there can be a mixture of them,

(2) there are three different types of immersions of leaves, namely the proper, (locally) dense and exceptional leaves. See I;4.1.2.

ii) Let $B = T^2$ and let $\pi_1(T^2,b_o)$ be generated by the two standard generators α and β. Let $G \subset \mathrm{Diff}^\infty_+(S^1)$ be the group of rotations of S^1 and denote by G^δ the group G considered as a discrete group. Let

$$H : \pi_1(T^2,b_o) \to G$$

be a representation. Then the following holds :

a) If $\xi_H = (M,p,T^2)$ is the suspension of H then ξ_H is trivial as a G-bundle but not as a G^δ-bundle if the representation H is non-trivial.

b) All leaves of ξ_H are dense in M if and only if $\Gamma = \text{im } H$ is dense in G.

c) All leaves of ξ_H are homeomorphic to $\mathbb{R}^2$ if and only if $H(\alpha)$ and $H(\beta)$ are irrational and rationally independent.

d) If Γ is dense in G then the leaves are homeomorphic either to $\mathbb{R}^2$ (the plane) or $S^1 \times \mathbb{R}$ (the cylinder).

e) If Γ is not dense in G then Γ is finite, all leaves of ξ_H are tori and together form the fibres of a fibre bundle with base S^1 and total space M.

f) In all cases above the holonomy of each leaf is trivial.

iii) Let $F = T^2$ and let Γ be a cyclic subgroup of the group T^2 with $g \in \Gamma$ as generator. For $B = S^1$, let

$$H : \pi_1 S^1 \to \Gamma$$
$$1 \mapsto g.$$

(Here $\pi_1 S^1$ is identified with $\mathbb{Z}$). There are different possibilities :

a) If Γ is finite then all leaves of ξ_H are homeomorphic to S^1 and they are the fibres of a fibre bundle with base T^2 and total space M.

b) If Γ is dense in T^2 then all leaves of ξ_H are homeomorphic to $\mathbb{R}$ and dense in M.

c) If $\bar{\Gamma}$ is a 1-dimensional subgroup of T^2 then the leaves are still homeomorphic to $\mathbb{R}$ and the closures of the leaves are the fibres of a fibre bundle with total space M, base S^1 and fibre T^2.

In all cases M is homeomorphic to the 3-torus.

1.2.14. - Exercises. i) Verify all assertions in the preceding examples.

ii) Let f be a homeomorphism of $I = [0,1]$ such that $f(x) < x$ for $x \in \overset{\circ}{I}$ and let $a \in \overset{\circ}{I}$ and $A = cl\{f^k(a)\}_{k\in\mathbb{Z}}$.

a) Find a homeomorphism g of I such that $f \circ g = g \circ f$ and A is the fixed point set of g.

Let $H : \pi_1 T^2 \to \mathrm{Homeo}(I)$ be a representation whose image Γ is generated by f and g. Let $\xi_H = (M,p,T^2)$ be the foliated bundle with holonomy H.

b) Show that M is homeomorphic to $T^2 \times I$, that $T^2 \times \{0\}$ and $T^2 \times \{1\}$ are leaves, that the leaf $\pi(\mathbb{R}^2 \times \{a\})$ is a cylinder (where $\pi : \mathbb{R}^2 \times I \to M$ is as in 1.2.1), and all other leaves are planes.

c) What are the holonomy groups of the leaves ?

d) Show that each leaf L is proper (i.e. the topology on L induced by the topology on $T^2 \times I$ equals the leaf topology on L).

Remark.- This type of example will be of importance to us (compare). It even can be made C^1 but not C^2, i.e. there is no homeomorphism h of I such that $f' = hfh^{-1}$ and $g' = hgh^{-1}$ are C^2 diffeomorphisms.

iii) This exercise deals with a particular case of Haefliger's theorem [Ha]; see chapter III for a full discussion.

Let $\xi_H = (M,p,B)$ be a foliated bundle with compact base B. Take $b_o \in B$ and put $F_o = p^{-1}(b_o)$.

a) Show that a leaf L is compact if and only if $L \cap F_o$ is finite.

b) Now suppose that $\operatorname{im} H$ is finite. Let $x_o \in F_o$. Show that there exists a neighbourhood W of x_o in F_o such that for every $x \in W$ the holonomy group Γ_x is contained in Γ_{x_o}.

c) Using b) show that for every $x \in W$ the leaf L_x is a covering of L_{x_o}.

1.3. Foliated bundles.

We have seen in 1.2.2 that the suspension of a representation $H : \pi_1(B,b_o) \to \text{Homeo}(F)$ provides a bundle $\xi_H = (M,p,B)$ with fibre F and discrete structure group $\text{Homeo}(F)^\delta$. This fibre bundle is defined up to S-isomorphism by the conjugacy class of H in $\text{Homeo}(F)$. As noticed in 1.2.12, i), all results of 1.2 remain true if $\text{Homeo}(F)$ is endowed with an arbitrary topology and $\text{Homeo}(F)^\delta$ is replaced by a totally disconnected subgroup $G \subset \text{Homeo}(F)$.

The purpose of this paragraph is to clarify the relation between suspensions and bundles with totally disconnected structure group. More precisely, we prove the following theorem which is a kind of converse to 1.2.2., ii) and 1.2.7.

1.3.1. - Theorem. - Let the group G be totally disconnected.

i) Every G-bundle $\xi = (M,p,B)$ with fibre F is G-isomorphic to the suspension with fibre F of a representation

$H_\xi : \pi_1(B,b_o) \to G$.

ii) Every G-bundle map (G-isomorphism) between suspensions of representations $H : \pi_1 B \to G$ and $H' : \pi_1 B' \to G$ is an S-map (resp. S-isomorphism).

We know that a cocycle corresponding to a G-atlas of a fibre bundle $\xi = (M,p,B)$ with totally disconnected structure group G is locally constant. We next want to analyze the special geometric properties of such a cocycle.

First let $C = (\{U_i\},\{g_{ij}\})$ be a cocycle on B with values in a (for the moment not necessarily totally disconnected) topological group. Recall from 1.1.11 the construction of the fibre bundle ξ_C with fibre F. As before, we denote by F^δ the set F provided with the discrete topology and by E^δ the disjoint union

$$E^\delta = \coprod_i (U_i \times F^\delta).$$

On E^δ we consider the equivalence relation ρ, as in 1.1.11., i.e.

$$(b,y)\,\rho\,(b',y') \quad \text{if} \quad b = b' \in U_i \cap U_j \quad \text{and} \quad y' = g_{ij}(b)(y).$$

1.3.2. - Lemma. - <u>The cocycle $C = (\{U_i\},\{g_{ij}\})$ on B is locally constant if and only if ρ is an open equivalence relation.</u>

Proof : For any g_{ij} consider the map

$$\begin{aligned} \phi_{ij} : (U_i \cap U_j) \times F^\delta &\longrightarrow (U_i \cap U_j) \times F^\delta \\ (b,y) &\longmapsto (b, g_{ij}(b)(y)). \end{aligned}$$

Then the following conditions are equivalent :

(1) g_{ij} is locally constant,

(2) ϕ_{ij} is a homeomorphism,

(3) ρ is open. □

For the rest of this paragraph the topological group G is supposed to be totally disconnected.

As in 1.2.3. we now introduce the leaf topology on $M = E/\rho$ which is induced by the quotient map $\pi : E^\delta \to M$. We use again M^δ as notation for M equipped with this topology.

1.3.3. - Lemma. - Let $\xi_C = (M,p,B)$ and $\xi_{C'} = (M',p',B')$ be G-bundles constructed by means of the (locally constant) cocycles C and C'.

i) $p : M^\delta \to B$ is a covering map.

ii) If $(f,\bar{f})$ is a G-bundle map (G-isomorphism) from ξ_C to $\xi_{C'}$ then we get a commutative diagram of continuous maps

$$\begin{array}{ccc} M^\delta & \xrightarrow{\ f\ } & M'^\delta \\ {\scriptstyle p}\downarrow & & \downarrow{\scriptstyle p'} \\ B & \xrightarrow{\ \bar{f}\ } & B' \end{array},$$

i.e. $(f,\bar{f})$ is a homomorphism (resp. an isomorphism) of covering maps.

Proof : Indeed, if $C = (\{U_i\},\{g_{ij}\})$ then $p^{-1}(U_i)$ is homeomorphic to $U_i \times F^\delta$, for every U_i, thus providing a trivialization of p. This shows that i) holds.

As G is totally disconnected the definition of a G-bundle map (resp. G-isomorphism if $B' = B$ and $\bar{f} = id$) immediately shows that $f : M \to M'$ is continuous for the leaf topologies. □

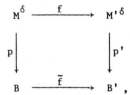

1.3.4. - Remark .- If in 1.3.3. the cocycles C and C' are equivalent then the proof shows that there is an isomorphism of coverings $f : M^{\delta} \to M'^{\delta}$, $p' \circ f = p$. As every G-bundle $\xi = (M,p,B)$ is G-isomorphic to some ξ_C this shows that the total space M carries a well-defined leaf topology, and 1.3.3. becomes true for arbitrary G-bundles.

1.3.5. - Definition.- Let $\xi = (M,p,B)$ be a G-bundle. By F_o, F_1 we denote the fibres over $b_o, b_1 \in B$, respectively. If $c : [0,1] \to B$ is a path from b_o to b_1 then, by 1.3.3. and 1.3.4., for every $y \in F_o$, there is a unique lifting $\tilde{c}_y$ of c from y to $\tilde{c}_y(1) \in F_1$. This defines a homeomorphism

$$T_c : F_o \to F_1$$

which is called the translation of F_o to F_1 along the path c. This map depends only on the homotopy class of c.

1.3.6. - Lemma.- Let $A = \{(U_i, \phi_i)\}$ be a G-atlas of $\xi = (M,p,B)$ and let (U_o, ϕ_o) and (U_1, ϕ_1) be local trivializations belonging to A. For $b_o \in U_o$ and $b_1 \in U_1$ the map

$$\phi_{1b_1} \circ T_c \circ \phi_{ob_o}^{-1} : F \to F$$

is a homeomorphism belonging to G.

Proof : It is clear that $\phi_{1b_1} \circ T_c \circ \phi_{ob_o}^{-1}$ is a homeomorphism. To see that it belongs to G we decompose it as a product of homeomorphisms belonging to G. We choose open sets $U_o = U_{j_o}, U_{j_1}, \ldots, U_{j_s} = U_1$ of $\{U_i\}$ and numbers $0 = t_o < t_1 < \ldots < t_{s+1} = 1$ such that $c([t_k, t_{k+1}]) \subset U_{j_k}$, $k = 0,\ldots,s$. Then we decompose T_c as the product of the translations $T_{c|[t_k,t_{k+1}]}$ using the homeomorphisms $\phi_{j_k b_k}$, with $b_k = c(t_k)$. □

1.3.7. - The holonomy representation. - Taking $b_1 = b_o$ in the above construction and fixing a local trivialization (U_o,ϕ_o) with $b_o \in U_o$, there is assigned to each $\gamma \in \pi_1(B,b_o)$ a homeomorphism $T_\gamma \in G$. Clearly, $T_{\gamma\gamma'} = T_{\gamma'} \circ T_\gamma$. Therefore, if we define H_ξ by $H_\xi(\gamma) = T_\gamma^{-1}$, then

$$H_\xi : \pi_1(B,b_o) \rightarrow G$$

is a homomorphism (which in general is neither injective nor surjective). It is called the holonomy representation of ξ. (Cf. Steenrod [St; p.61]. There H_ξ is called a characteristic class). It is defined up to conjugation with an element of G depending on the choice of (U_o,ϕ_o).

As a first result we obtain:

1.3.8. - Lemma. - A fibre bundle $\xi = (M,p,B)$ with fibre F and totally disconnected structure group G is trivial if and only if its holonomy representation H_ξ is trivial.

Proof : If ξ is trivial then evidently H_ξ is also trivial. So let us suppose that $H_\xi : \pi_1(B,b_o) \rightarrow G$ is trivial. Then for every path c in B starting in b and ending in b_o the translation $T_c : F_b \rightarrow F_{b_o}$ is independent of c. We denote it by T_b. After identification of F_{b_o} with F by ϕ_{ob_o}, as in 1.3.6.,

$$\begin{aligned} h : M &\longrightarrow B \times F \\ x &\longmapsto (p(x), T_{p(x)}(x)) \end{aligned}$$

is a G-isomorphism. □

We are now ready to prove theorem 1.3.1.

Proof of theorem 1.3.1.- Let $\tilde{\xi} = (\tilde{M},\tilde{p},\tilde{B})$ be the G-bundle induced from ξ by the universal covering map $q : \tilde{B} \to B$, that is we have the commutative diagram

$$(*) \qquad \begin{array}{ccc} \tilde{M} & \xrightarrow{\tilde{p}} & \tilde{B} \\ {\scriptstyle \tilde{\pi}}\downarrow & & \downarrow{\scriptstyle q} \\ M & \xrightarrow{p} & B \end{array} .$$

As $\tilde{B}$ is simply connected we have $H_{\tilde{\xi}} \equiv id$ and thus, by 1.3.8., $\tilde{\xi}$ is G-isomorphic to the product bundle $(\tilde{B} \times F, pr, \tilde{B})$. Using this isomorphism the diagram (*) becomes

$$\begin{array}{ccc} \tilde{B} \times F & \xrightarrow{pr} & \tilde{B} \\ {\scriptstyle \pi}\downarrow & & \downarrow{\scriptstyle q} \\ M & \xrightarrow{p} & B \end{array} ,$$

where (π,q) is a G-bundle map, see also 1.1.11. Using 1.3.3., ii) it is not hard to see that π is a regular covering with group of covering translations equal to $\pi_1 B$ acting on $\tilde{B} \times F$ by

$$\begin{array}{rcl} A : \pi_1 B \times (\tilde{B} \times F) & \to & \tilde{B} \times F \\ (\gamma,(\tilde{b},y)) & \mapsto & (\gamma(\tilde{b}),H_{\xi}(\gamma)(y)). \end{array}$$

This proves i).

For the second part, consider the diagram

$$\begin{array}{ccc} M & \xrightarrow{f} & M' \\ {\scriptstyle p}\downarrow & & \downarrow{\scriptstyle p'} \\ (B,b_o) & \xrightarrow{\bar{f}} & (B',b_o') \end{array}$$

and local trivializations (U_o,ϕ_o) and (U'_o,ϕ'_o) of ξ_H and $\xi_{H'}$, respectively, with $b_o \in U_o$ and $b'_o \in U'_o$. With the notations of 1.1.7., the homeomorphism

$$g = \phi'_{ob'_o} \circ f_{b_o} \circ \phi^{-1}_{ob_o}$$

is an element of G. Using 1.3.3., ii) (in connection with the remark following it) we conclude that

$$g\, H_\xi(\gamma) = H_{\xi'}(\bar{f}_\# \gamma) g$$

for every $\gamma \in \pi_1(B,b_o)$. Thus applying 1.2.8. we see that $(f,\bar{f})$ is an S-map, or an S-isomorphism when f is a G-isomorphism, as required. □

1.3.9. - Remark. - By the preceding discussion, we need no longer distinguish between suspensions and fibre bundles with totally disconnected structure group. Also it is justified to call these objects foliated bundles.

In future the notions "suspension" and "foliated bundle" will be used synonymously.

1.3.10. - Remark.- Let $\xi = (M,p,B)$ be a fibre bundle with (not necessarily totally disconnected) structure group G. Suppose that G can be reduced to the totally disconnected subgroups Γ_1 and Γ_2 of G. Denote by ξ_i the Γ_i-bundle ξ. Both ξ_1 and ξ_2 are then foliated bundles. It may happen, even if $\Gamma_1 = \Gamma_2$, that these two fibre bundles are not S-isomorphic, that is there is no homeomorphism $f : M \to M$ taking the leaves of ξ_1 onto the leaves of ξ_2. (See exercise 1.3.12. i) and also 1.2.13. ii)).

1.3.11. - Example. - Let M be an m-dimensional affine manifold, that is M admits an atlas $\mathcal{A} = \{(U_j,\varphi_j)\}$ such that each coordinate change $\varphi_{ij} = \varphi_i \circ \varphi_j^{-1}|\varphi_j(U_i \cap U_j)$ is the restriction of an affine motion $A_{ij} + v_{ij}$, with $A_{ij} \in GL(m;\mathbb{R})$ and $v_{ij} \in \mathbb{R}^m$. Thus if $A(m;\mathbb{R})$ denotes the group of affine motions of $\mathbb{R}^m$ then to $\mathcal{A}$ there are associated two locally constant cocycles

$C = (\{U_j\},\{g_{ij}\})$ with $g_{ij} : U_i \cap U_j \to A(m;\mathbb{R})$ defined by $g_{ij}(x) = A_{ij} + v_{ij}$, and

$C' = (\{U_j\},\{g'_{ij}\})$ with $g'_{ij} : U_i \cap U_j \to GL(m;\mathbb{R})$ defined by $g'_{ij}(x) = A_{ij}$.

Thus C and C' yield foliated bundles over M with fibre $\mathbb{R}^m$.

1.3.12.- Exercises. i) Let $B = S^1$ and $G \subset \mathrm{Diff}^\infty(S^1)$ the group of rotations.

a) Show that the suspensions ξ_α and ξ_β of any two rotations $\bar{R}_\alpha$ and $\bar{R}_\beta$, respectively, are G-isomorphic.

b) Find a pair of real numbers (α,β) such that ξ_α and ξ_β are not S-isomorphic.

c) If Γ denotes the subgroup of G generated by $\bar{R}_\alpha$ and $\bar{R}_\beta$ under what condition are ξ_α and ξ_β Γ-isomorphic ?

ii) Let $\xi = (S^3,p,S^2)$ be the Hopf bundle (fibre S^1 and structure group S^1). Show that S^1 cannot be reduced to a totally disconnected subgroup.

1.4. Equivariant submersions.

1.4.1. - As we have seen in the last section a foliated bundle $\xi = (M,p,B)$ with fibre F is characterized by the existence of a regular covering $\pi : \tilde{B} \times F^\delta \to M^\delta$ whose group of covering translations preserves the product structure of $\tilde{M} = \tilde{B} \times F$. If $D : \tilde{M} \to F$ is the canonical projection then we have the diagram

$$(*) \qquad \begin{array}{ccc} \tilde{M} & \xrightarrow{D} & F \\ {\scriptstyle\pi}\big\downarrow & & \\ M & & \end{array}$$

On our way to the general notion of foliations we are going to weaken this somewhat rigid structure in the following way. Roughly speaking we forget the bundle structure on M but not the leaf topology. To be more precise we consider a diagram (*) where

(1) π is a regular covering and

(2) D is a submersion equivariant under the group Γ of covering translations, i.e. each $\gamma \in \Gamma$ takes the fibres of D onto fibres.

A diagram (*) satisfying (1) and (2) is called an <u>equivariant submersion</u>. Note that $\tilde{M}$ need no longer be a product and that D is not assumed to be surjective. (In the C^0 case a <u>submersion</u> $D : M^m \to F^n$, $n \leq m$, is locally of the form $D = pr \circ \varphi$, where φ is a homeomorphism in $\mathbb{R}^m$ and $pr : \mathbb{R}^m = \mathbb{R}^{m-n} \times \mathbb{R}^n \to \mathbb{R}^n$ is the canonical projection).

The action of Γ on $\tilde{M}$ induces an action of Γ on $\operatorname{im} D$, but - as we shall see (example 1.4.4.) and already have seen with foliated bundles - this action in general is not properly discontinuous and the quotient $(\operatorname{im} D)/\Gamma$ is not Hausdorff.

The <u>leaf topology</u> on $\tilde{M}$ is by definition generated by the intersections of open sets of $\tilde{M}$ with the fibres of D. Clearly, the components of this topology are the components of the fibres of D. We call them leaves and denote by $\tilde{M}^\delta$ the set $\tilde{M}$ provided with the leaf topology, in accordance with the last paragraph. By the condition (2), the group Γ acts also continuously on $\tilde{M}^{\delta}$ and, passing to the quotient, we get a topology on M which is again called the leaf topology. If M^{δ} denotes M supplied with this topology then $\pi : \tilde{M}^{\delta} \to M^{\delta}$ becomes a covering map.

1.4.2. - *Definition*. - We say that, by the above construction, there is on M a foliation obtained by an equivariant submersion.

In order to clarify this notion we observe the following situation.

1.4.3. - *Lemma* (of local trivialization).- Let M and F be manifolds of dimension m and n, respectively. Given an equivariant submersion

$$\begin{array}{ccc} \tilde{M} & \xrightarrow{D} & F \\ {\scriptstyle\pi}\downarrow & & \\ M & & \end{array}$$

there is

(1) an open covering $\mathcal{U}$ of M,

(2) a submersion $f_i : U_i \to \mathbb{R}^n$, for every $U_i \in \mathcal{U}$,

(3) a homeomorphism $\gamma_{ij} : f_j(U_i \cap U_j) \to f_i(U_i \cap U_j)$ such that

$$f_i = \gamma_{ij} \circ f_j$$

for every pair U_i, U_j with $U_i \cap U_j \neq \emptyset$.

Proof : Using the restrictions of the submersion D on open sets of $\tilde{M}$ it is easily seen that on $\tilde{M}$ there are maps satisfying (1) - (3). As D is equivariant we may take an open covering $\{U_j\}$

of M over which π is trivialized to define maps f_j and γ_{ij} such that $\{(U_j,f_j)\},\{\gamma_{ij}\}$ fulfill conditions (1) - (3). We omit the details. □

Before we conclude our preparations with an example let us remark that - as for foliated bundles - the notion of holonomy applies also to a foliation obtained by an equivariant submersion $M \xleftarrow{\pi} \tilde{M} \xrightarrow{D} F$.

As was noticed earlier the group of covering translations Γ acts also on $\mathrm{im}(D)$. Therefore, as with foliated bundles, we get a homomorphism

$$H : \pi_1(M,x) \to \mathrm{Homeo}(\mathrm{im}\, D).$$

Let $\tilde{x} \in \pi^{-1}(x)$, $y = D(\tilde{x})$ and L_x the leaf passing through x. We denote by Γ_y the isotropy group of Γ in $y \in F$ and get the commutative diagram (see also 1.2.10) :

$$\begin{array}{ccc} \pi_1(L_x,x) & \xrightarrow{H_y} & \Gamma_y \\ {\scriptstyle i_\#}\big\downarrow & & \big\downarrow \\ \pi_1(M,x) & \xrightarrow{H} & \Gamma \end{array}$$

The representation H_y is again called the <u>holonomy representation of the leaf</u> L_x <u>in</u> x. (We shall come back to the concept of holonomy in the general case in chapter III, for example).

1.4.4. - Reeb components. - We consider the following example of a foliation obtained by an equivariant submersion.

Let $\tilde{M} = \mathbb{R}^m_* = \mathbb{R}^m - \{0\}$ and $D : \tilde{M} \to \mathbb{R}$ be the projection on the m-th coordinate. On $\tilde{M}$ we define an action of $\mathbb{Z}$ which is generated by the diffeomorphism

$$\varphi : \tilde{M} \to \tilde{M}, \quad x \mapsto \frac{1}{2} x.$$

Clearly, D is equivariant under this action. The quotient map

$$\pi : \tilde{M} \to M = \tilde{M}/\mathbb{Z}$$

is an infinite cyclic covering. By looking at the fundamental domain

$$K = \{(x_1,\ldots,x_m) \in \mathbb{R}^m \mid \quad 1 \leq \sqrt{x_1^2 + \ldots + x_m^2} \leq 2\},$$

which is indicated for $m = 2$ in fig. 2, we see that $\mathbb{R}^m_*/\mathbb{Z}$ is homeomorphic to $S^1 \times S^{m-1}$.

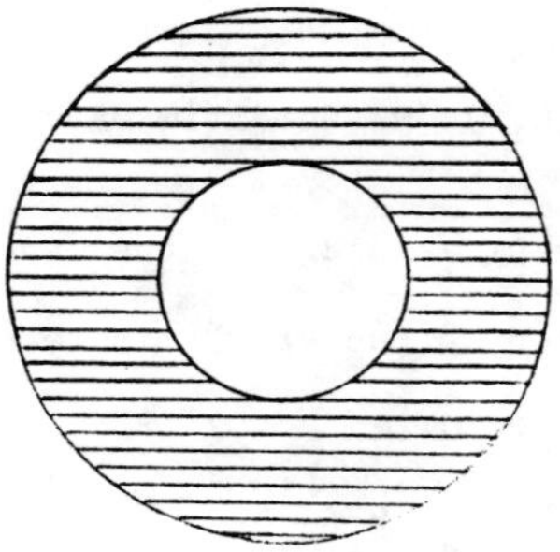

Figure 2

We thus get on $S^1 \times S^{m-1}$ a foliation obtained by the submersion $D = pr_m$ which is equivariant with respect to the action of $\mathbb{Z}$ on $\mathbb{R}^m_*$ generated by $\emptyset$. We call it a Reeb foliation on $S^1 \times S^{m-1}$.

This construction was communicated to the authors by B. Anderson.

If we restrict the maps D and π to $\mathbb{H}^m_* = \{x \in \mathbb{R}^m_* \mid D(x) \geq 0\}$ then we get a quotient space $\mathbb{H}^m_*/\mathbb{Z}$ which is homeomorphic to $S^1 \times D^{m-1}$.

1.4.5. - Definition.- The foliation on $\mathbb{H}^m_*/\mathbb{Z} = S^1 \times D^{m-1}$ obtained by the restricted equivariant submersion $D|\mathbb{H}^m_*$ is called a (m-dimensional) Reeb component (or Reeb foliation on $S^1 \times D^{m-1}$).

The leaves of a Reeb component consist of the boundary $S^1 \times S^{m-2}$ and the images $\pi(\mathbb{R}^{m-1} \times \{x_m\}), x_m > 0$, which are all homeomorphic to $\mathbb{R}^{m-1}$ and are embedded submanifolds.

1.4.6. - Exercises. i) Show that, for $m = 2$, a Reeb component, as defined in 1.4.5, is also a Reeb component in the sense of I; 3.3.1.

ii) What are the holonomy representations of the leaves of the Reeb foliation on $S^1 \times S^{m-1}$ or of the Reeb component defined above?

2. *Foliated manifolds.*

2.1. *Definition of a foliation ; related notions.*

We have arrived at the central objective of our investigations, the foliations on arbitrary manifolds. Let us begin with an intuitive description of what is without doubt the best known example, the Reeb foliation on the 3-sphere S^3 .

We may think of S^3 as the union of two solid tori $D^2 \times S^1$ which are glued together by means of a homeomorphism on the boundary that interchanges meridian and longitude. If each of these solid tori is considered as a Reeb component (see 1.4.5) then the partition of S^3 by surfaces we get in this way is called the Reeb foliation on S^3. It was found by [Re]. This foliation cannot be defined by an equivariant submersion.

2.1.1. - Definition.- Let M be an m-dimensional manifold ‡ without boundary, $n \leqslant m$ and $0 \leqslant r \leqslant \infty$ or $r = \omega$ (real analytic).

i) A C^r atlas $A = \{(U_i, \varphi_i)\}$ on M is called a foliated C^r atlas of codimension n if the coordinate transformations

$$\varphi_{ij} = \varphi_i \circ \varphi_j^{-1} | \varphi_j(U_i \cap U_j) \to \mathbb{R}^m = \mathbb{R}^{m-n} \times \mathbb{R}^n$$

are of the form

(*) $$\varphi_{ij}(x_1,\ldots,x_m) = (\alpha_{ij}(x_1,\ldots,x_m),\ \gamma_{ij}(x_{m-n+1},\ldots,x_m)) .$$

‡ Manifolds are assumed to be connected and with a countable basis, unless otherwise stated.

ii) Two foliated C^r atlases of codimension n on M are <u>equivalent</u> if their union is again a foliated C^r atlas of codimension n.

Note that any refinement $A' = \{(U'_\kappa, \varphi'_\kappa)\}$ of a foliated C^r atlas $A = \{(U_i, \varphi_i)\}$ of codimension n is again a foliated C^r atlas of codimension n, and A and A' are equivalent.

iii) The m-manifold M together with an equivalence class F of foliated C^r atlases of codimension n (maximal foliated atlas) is called a <u>foliated manifold of class</u> C^r <u>and codimension</u> n (or <u>dimension</u> $\ell = m-n$).

We use (M,F) as notation, or simply F when there is no doubt which manifold is meant ; the terminology "F <u>is a</u> (C^r) <u>foliation</u> (<u>on</u> M)" will also be used frequently.

iv) If $\partial M \neq \emptyset$ then the definition of a foliated atlas $A = \{(U_i, \varphi_i)\}$ has to be modified in such a way that for each boundary component M_o of M and $U_i \cap M_o \neq \emptyset$ we have $\varphi_i : U_i \to \mathbb{H}^m$, where either

$$\mathbb{H}^m = \mathbb{H}^m_1 = \{(x_1,\ldots,x_m) \in \mathbb{R}^m | x_1 \geq 0\}$$

or

$$\mathbb{H}^m = \mathbb{H}^m_m = \{(x_1,\ldots,x_m) \in \mathbb{R}^m | x_m \geq 0\}.$$

Note that it follows from (*) that two charts (U_i, φ_i) and (U_j, φ_j) such that $U_i \cap M_o \neq \emptyset$ and $U_j \cap M_o \neq \emptyset$ both have image in $\mathbb{H}^m_1$ or both have image in $\mathbb{H}^m_m$.

In the first case we say that the foliation is <u>transverse</u> to M_o and in the second case we say the foliation is <u>tangent</u> to M_o.

v) In the definition of a foliation of course we could have used any other factorization originally, but having chosen one we have to stick to it, unless we add to the notation an indication of the chosen factorization of $\mathbb{R}^m$.

2.1.2. - *Basic examples*. - i) Every C^r structure of an m-manifold M may be considered as a C^r foliation of codimension 0 or of codimension m.

ii) $\mathbb{R}^m$ together with the C^r foliation given by the trivial atlas consisting of the single chart $(\mathbb{R}^m, id_{\mathbb{R}^m})$ is a foliated manifold of class C^r of codimension n, for any $n \leqslant m$ and any r, $0 \leqslant r \leqslant \omega$.

iii) Every C^r vector field without singularities on M yields a 1-dimensional foliation of class C^r on M.

A foliation can be defined in a slightly different way by means of a "foliated cocycle", a notion which on the one hand is similar to that of a cocycle as used in § 1 but on the other hand is more rigid. For this we need the following definition of a pseudogroup, a concept we shall apply in many cases in the subsequent chapters. (For simplicity, we consider here only manifolds without boundary. The modifications necessary for manifolds with boundary are straightforward.)

2.1.3. - *Definition*. i) By a pseudogroup of local homeomorphisms (diffeomorphisms, etc.) we mean a family $H = \{h_i : D_i \to R_i\}$ of homeomorphisms (diffeomorphisms, etc.) between open subsets of a topological space T such that the following conditions are fulfilled :

(1) If h_i belongs to H then $h_i^{-1} : R_i \to D_i$ belongs to H.

(2) If $h_i, h_j \in H$ and $D_i \cap R_j \neq \emptyset$ then $h_i \circ h_j : h_j^{-1}(D_i \cap R_j) \to h_i(D_i \cap R_j)$ belongs to H.

(3) $id_T \in H$.

Usually, there is included a forth axiom in the definition of a pseudogroup (see Sacksteder [Sa], for instance). This axiom, however, is of no relevance to our purposes :

(4) If $h_i, h_j \in H$, $D_i \cap D_j \neq \emptyset$ and $h : D_i \cup D_j \to R_i \cup R_j$ is a homeomorphism with $h(x) = h_i(x)$ for $x \in D_i$ and $h(x) = h_j(x)$ for

$x \in D_j$ then $h \in H$.

ii) A <u>sub-pseudogroup</u> H_o of H is a subset of H which simultaneously is a pseudogroup.

For $0 \leqslant r \leqslant \omega$, we denote by H_n^r the pseudogroup of local C^r diffeomorphisms of $\mathbb{R}^n$. Clearly, H_n^r contains as sub-pseudogroup the pseudogroups H_n^s, $s > r$, and H_{n+}^r of orientation preserving local C^r diffeomorphisms of $\mathbb{R}^n$.

We shall also speak of local submersions or, more generally, of <u>local maps</u>, i.e. maps which are defined only on an open subset. Recall that a C^s map $f : M \to N$ between C^r manifolds M and N, $s \leqslant r$, is a C^s <u>submersion</u> if either $s \geqslant 1$ and the tangential map $T_x f$ is surjective, for every $x \in M$, or $s = 0$ and f is locally given by $f = \psi^{-1} \circ pr_{mn} \circ \varphi$, where $m = \dim M \geqslant n$, φ and ψ are charts on M and N, respectively, and $pr_{mn} : \mathbb{R}^m = \mathbb{R}^{m-n} \times \mathbb{R}^n \to \mathbb{R}^n$ is the canonical projection.

2.1.4. - *<u>Definition</u>*.- Let M be an m-dimensional manifold and $n \leqslant m$. By a <u>foliated cocycle on</u> M <u>with values in</u> H_n^r we mean a pair $C = (\{(U_i,f_i)\},\{g_{ij}\})$ where $\{U_i\}$ is an open covering of M, $f_i : U_i \to \mathbb{R}^n$ is a C^r submersion, for every i, and when $U_i \cap U_j \neq \emptyset$ the maps

$$g_{ij} : U_i \cap U_j \to H_n^r$$

are locally constant and satisfy

$$f_i(x) = g_{ij}(x)(f_j(x)) \text{ , for every } x \in U_i \cap U_j \text{ .}$$

Note that for $x \in U_i \cap U_j \cap U_k$ the cocycle condition

$$g_{ik}(x) = g_{ij}(x) \circ g_{jk}(x)$$

holds in a neighbourhood of $f_k(x) \in \mathbb{R}^n$.

We next want to see what relation there is between the foliated atlases and the foliated cocycles on M.

2.1.5. - Every foliated C^r atlas $A = \{(U_i, \varphi_i)\}$ of codimension n of M determines a foliated cocycle C on M with values in H^r_n. If $\varphi_i \circ \varphi_j^{-1} = (\alpha_{ij}, \gamma_{ij})$ we put $f_i = pr_{mn} \circ \varphi_i$ and define $g_{ij} : U_i \cap U_j \to H^r_n$ by $g_{ij}(x) = \gamma_{ij}$. Then C is given by $(\{(U_i, f_i)\}, \{g_{ij}\})$. We call C the foliated cocycle <u>corresponding</u> to A.

Conversely, any foliated cocycle $C = (\{(U_i, f_i)\}, \{g_{ij}\})$ on M with values in H^r_n yields a foliated C^r atlas of codimension n of M in the following way.

The C^r submersion $f_i : U_i \to \mathbb{R}^n$ can locally be written as

$$f_i = pr_{mn} \circ \varphi_i,$$

where φ_i is a local C^r diffeomorphism. This follows for $r = 0$ by the definition of a C^0 submersion and for $r \geq 1$ by the implicit function theorem. Hence there is a refinement $\{V_\kappa\}$ of $\{U_i\}$ and local C^r diffeomorphism $\psi_\kappa : V_\kappa \to \mathbb{R}^m$ which are of the form $\psi_\kappa = \varphi_i | V_\kappa$ for some φ_i. If $V_\kappa \cap V_\lambda \neq \emptyset$ then in a neighbourhood of $x \in V_\kappa \cap V_\lambda$ one has the commutative diagram of local maps

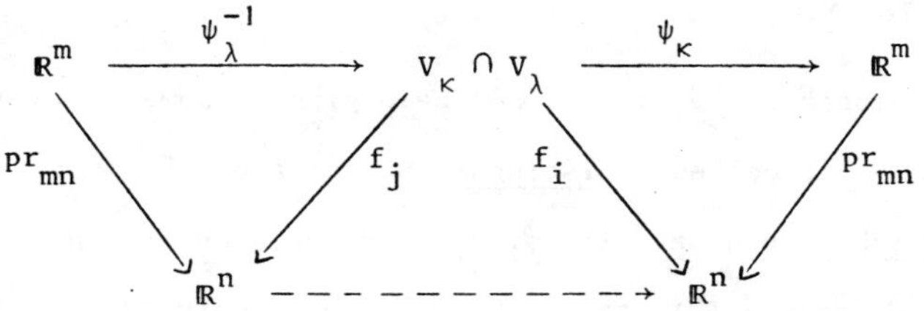

where $\gamma_{\kappa\lambda} = g_{ij}(x)$, in a neighbourhood of $f_j(x)$. This shows that the coordinate transformations $\psi_\kappa \circ \psi_\lambda^{-1}$ are of the form (*) in 2.1.1.

Thus by $A = \{(V_\kappa, \psi_\kappa)\}$ we have obtained from C a foliated C^r atlas of codimension n of M.

Evidently, A is not uniquely determined by the foliated cocycle C. This is due to the fact that the maps ψ_κ are not uniquely determined by the submersions f_i. But clearly any two foliated atlases constructed in the above manner by means of the foliated cocycle C are equivalent. Thus if we call two foliated cocycles on M with values in H^r_n equivalent if and only if their corresponding foliated atlases are equivalent then we get a bijection between

(1) the set of equivalence classes of foliated C^r atlases of codimension n of M and

(2) the set of equivalence classes of foliated cocycles on M with values in H^r_n.

Therefore, a foliation on M is given either by a foliated C^r atlas or by a foliated cocycle.

We adapt some notations of the first chapter to the more general situation at hand.

2.1.6. - Definition. - Let F be a foliation of codimension n on the m-manifold M.

i) A homeomorphism $\varphi_i : U_i \to \mathbb{R}^m$ belonging to some foliated atlas representing F is called a distinguished chart of F, the set U_i a distinguished open set. If $\varphi_i(U_i)$ is the open (resp. closed) m-cube $\mathring{I}^m$ (resp. I^m) then (U_i, φ_i), or simply U_i, is called a distinguished open (resp. closed) m-cube of F. (With slight modifications this notion applies also to foliated manifolds with boundary).

Note that every $x \in M$ is contained in an open (or closed) distinguished m-cube of F.

ii) The submersions $f_i = pr_{mn} \circ \varphi_i : U_i \to \mathbb{R}^n$, where (U_i, φ_i) is a distinguished chart of f, are referred to as <u>distinguished maps</u> of F. We use (U,f) as notation.

iii) Let $f_i : U_i \to \mathbb{R}^n$ be a distinguished map of F. The components of the fibres of f_i are called <u>plaques</u> of F.

Note that the intersection of two plaques is a union of plaques.

iv) The plaques of F for the different distinguished maps of F form a basis of a topology on the set M which makes M an ℓ-dimensional manifold, $\ell = m-n$. This topology is called the <u>leaf topology</u> of (M,F). We shall use the notation M^δ for M endowed with the leaf topology.

v) The components of M^δ are called the <u>leaves</u> of F. The leaves are injectively immersed ℓ-dimensional submanifolds of M. Through each point $x \in M$ there passes exactly one leaf of F which is usually denoted by L_x. We continue to write $L \in F$ for "L is a leaf of F".

vi) A <u>homomorphism of foliated manifolds</u> (an F-<u>homomorphism</u>, for short)

$$f : (M,F) \to (M',F')$$

is a map $f : M \to M'$ which is continuous for both the manifold topologies and the leaf topologies of M and M'. (This implies that the leaves of F are mapped into leaves of F'.)

The notion of an isomorphism between foliated manifolds (F-isomorphism) is then introduced in the obvious way. Note that an F-homomorphism or F-isomorphism $f : (M,F) \to (M',F')$ can be of a lower differentiability class than that of F and F'. Isomorphic foliations are also said to be conjugate or homeomorphic. Often we shall also refer to the differentiability class of an F-isomorphism.

vii) A subset A of the foliated manifold (M,F) is called saturated if it is a union of leaves. The empty set is by definition saturated.

If A is a saturated or any open submanifold of (M,F) then any foliated atlas of (M,F) induces a foliated atlas of A ; the foliation $F|A$ on A given in this way is called the foliation induced by F on A or the restriction of F to A .

Note that the inclusion

$$i_A : (A,F|A) \to (M,F)$$

is a homomorphism (of foliated manifolds). We shall see later (2.2.) that under certain circumstances F induces also a foliation on non-saturated submanifolds .

2.1.7. - Further examples. - i) Every C^r submersion $f : M \to N$ of the m-manifold M onto the n-manifold N gives rise to a C^r foliation of codimension n on M. A foliated cocycle on M with values in H^r_n is easily constructed by means of f and any C^r atlas of N.

In particular, if $M = B \times F$ then the two natural projections pr_B and pr_F define foliations on M whose leaves are of the form $\{b\} \times F$ or $B \times \{y\}$, respectively. In the first case the foliation is called the vertical foliation, in the second case the horizontal foliation on M.

A foliation globally defined by a submersion is sometimes called <u>simple</u>.

ii) Given a foliated bundle $\xi = (M,p,B)$ with fibre F and totally disconnected structure group G, a foliated cocycle on M can be constructed as follows. Let $\mathcal{A} = \{(U_i,\phi_i)\}$ be a G-atlas of ξ and let $C = (\{U_i\},\{g_{ij}\})$ be the cocycle corresponding to $\mathcal{A}$ (in the sense of 1.1.4.). Then the map

$$f_i = pr_F \circ \phi_i : p^{-1}(U_i) \to F$$

is a submersion.

Now we take an atlas $\mathcal{A}_F = \{(V_\kappa,\psi_\kappa)\}$ of F and denote by $\mathcal{W}$ the family of open sets W_α of M of the form

$$W_\alpha = p^{-1}(U_i) \cap f_i(V_\kappa),$$

where U_i runs through $\{U_i\}$ and V_κ runs through $\{V_\kappa\}$.

If $n = \dim F$ then for each $W_\alpha \in \mathcal{W}$ the map

$$f'_\alpha = \psi_\kappa \circ (f_i|W_\alpha) : W_\alpha \to \mathbb{R}^n$$

is a local submersion.

Finally, if $W_\alpha = p^{-1}(U_i) \cap f_i(V_\kappa)$ and $W_\beta = p^{-1}(U_j) \cap f_j(V_\lambda)$ have a non-empty intersection then for every $x \in W_\alpha \cap W_\beta$ there is a commutative diagram

$$\begin{array}{ccccc}
 & & V_\kappa & \xrightarrow{\psi_\kappa} & \mathbb{R}^n \\
 & \nearrow^{f'_\alpha} & & & \\
W_\alpha \cap W_\beta & & \downarrow g_{ij}(b) & & \downarrow g'_{\alpha\beta}(x) \\
 & \searrow_{f'_\beta} & & & \\
 & & V_\lambda & \xrightarrow{\psi_\lambda} & \mathbb{R}^n
\end{array}$$

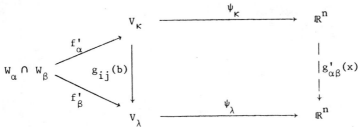

where $b = p(x)$. The fact that the cocycle $\{g_{ij}\}$ is locally constant implies that $g'_{\alpha\beta} : W_\alpha \cap W_\beta \to H^r_n$ defined by the diagram is locally constant and $\{g'_{\alpha\beta}\}$ together with the family $\{(W_\alpha, f'_\alpha)\}$ fulfills the conditions (1) and (2) of 2.1.4.

iii) As was already indicated in 1.4.3, and will be stated explicitly as corollary 2.2.6, any equivariant submersion $M \xleftarrow{\pi} \tilde{M} \xrightarrow{D} F^n$ yields a foliation of codimension n on M. We conclude that the Reeb component on $S^1 \times D^{m-1}$ and the Reeb foliation on $S^1 \times S^{m-1}$, as defined in 1.4.5, are real analytic foliations of codimension one.

2.1.8. - Exercises.- i) Give a condition for foliated cocycles to be equivalent similar to that of 1.1.6.

ii) Show that the Reeb foliation on S^3, as described at the beginning of this paragraph, is a C^0 foliation of codimension one.

iii) Let $(\tilde{f}, f, \bar{f})$ be a map (resp. isomorphism) of suspensions $\xi_H = (M,p,B)$ and $\xi_{H'} = (M',p',B')$, cf. 1.2.6. Show that f is a homomorphism (isomorphism) of the corresponding foliated manifolds $(M,\mathcal{F})$ and $(M',\mathcal{F}')$. Show that not every $\mathcal{F}$-homomorphism $f : (M,\mathcal{F}) \to (M',\mathcal{F}')$ is a map of suspensions. Give an example.

iv) Show that the leaves of a foliation are manifolds whose topologies have countable bases.

v) (cf. I; 4.1). Let A be a saturated subset of $(M,\mathcal{F})$. Denote by $\operatorname{int} A$ the interior of A in M. Show that $\bar{A}$, $\operatorname{int} A$, $\operatorname{int} \bar{A}$, $\bar{A} - \operatorname{int} \bar{A}$ are saturated.

2.2. *Transversality; orientability.*

2.2.1. - *Remarks and definitions.* - Let (M,F) be a foliated manifold of codimension n and M' a manifold (without boundary).

i) A map $f : M' \to M$ is transverse to F in the point $x' \in M'$ if there is a neighbourhood U' of x' and a distinguished chart (U,g) of F, $f(U') \subset U$, such that

$$g \circ f : U' \to \mathbb{R}^n$$

is a submersion.

ii) The map f is transverse to F if f is transverse to F in each $x' \in M'$. For example, if $(f,\bar{f}) : \xi \to \xi'$ is a homomorphism of suspensions then $f : M \to M'$ is transverse to the corresponding foliation on M'.

iii) A submanifold M' of M is called transverse to F if the inclusion $i_{M'} : M' \to M$ is transverse to F. Note that if $\partial M \neq \emptyset$ and $M_o \subset \partial M$ is transverse to F then M_o is transverse to F in the sense of 2.1.1. and vice versa.

A foliation F' is said to be transverse to F if each leaf of F' is transverse to F.

iv) It is not hard to see that if F' is transverse to F then F is also transverse to F'. In this case we may introduce bidistinguished cubes (with respect to F and F'), as in chapter I. It is clear that the bidistinguished cubes form a basis of the topology on M.

v) In contrast to the case of foliated surfaces, a foliation in general does not admit any transverse foliation, (cf. the next exercises and 2.3).

vi) Finally, if M' is a manifold with boundary we say that $f : M' \to (M,\mathcal{F})$ is transverse to $\mathcal{F}$ if both $f|\partial M'$ and $f|\mathring{M}'$ are transverse to $\mathcal{F}$.

2.2.2. - *Lemma*.- Let $f : M' \to (M,\mathcal{F})$ be a transverse map, where $\mathcal{F}$ is of codimension n. There exists a unique foliation $\mathcal{F}'$ of codimension n on M' such that

$$f : (M',\mathcal{F}') \to (M,\mathcal{F})$$

is a homomorphism of foliations.

If f and $\mathcal{F}$ are of class C^r then so is $\mathcal{F}'$.

Proof : We construct a suitable foliated cocycle on M' with values in $\mathcal{H}^r_n$. Let $C = (\{(U_i,f_i)\},\{g_{ij}\})$ be a foliated cocycle on M and $\{U'_\kappa\}$ an open cover of M' subordinate to $\{f^{-1}(U_i)\}$. For $U_\kappa \subset f^{-1}(U_i)$ the map

$$f'_\kappa = f_i \circ f : U_\kappa \to \mathbb{R}^n$$

is a local C^r submersion. For $U'_\kappa \cap U'_\lambda \neq \emptyset$, $U'_\lambda \subset f^{-1}(U_j)$, we define

$$g'_{\kappa\lambda} : U'_\kappa \cap U'_\lambda \longrightarrow \mathcal{H}^r_n$$

by $g'_{\kappa\lambda}(x') = g_{ij}(x)$, where $x = f(x')$. Then $C' = (\{(U'_\kappa,f'_\kappa)\},\{g'_{\kappa\lambda}\})$ is a foliated cocycle on M' with values in $\mathcal{H}^r_n$. If $\mathcal{F}'$ is the foliation given by C' then by construction

$$f : (M',\mathcal{F}') \longrightarrow (M,\mathcal{F})$$

is an $\mathcal{F}$-homomorphism.

The uniqueness of $\mathcal{F}'$ follows from the fact that the leaves of $\mathcal{F}'$ are the components of the inverse images $f^{-1}(L)$, for $L \in \mathcal{F}$. □

2.2.3. - *Definition*.- The foliation F' on M' obtained by 2.2.2. is called the foliation <u>induced by</u> f <u>on</u> M (or the <u>lift</u> of F by f) ; it is denoted by f^*F.

In particular, if M' is a submanifold transverse to F then the foliation $i_{M'}^*F$ is said to be the foliation <u>induced by</u> F <u>on</u> M' ; it is usually denoted by $F|M'$. This is in accordance with the notion of the induced foliation on a saturated submanifold, as introduced in 2.1.6., vii). Note that if $M' \subset M$ is open then $F|M'$ is always defined.

The previous situation obviously applies when the map f is a covering map

$$\pi : M' \to (M,F).$$

Evidently, π is transverse to F and thus we have the foliation π^*F on M'.

One could ask conversely under what conditions a foliation F' on M' projects under π to a foliation on M. Here is the answer to this question for π a regular covering.

2.2.4. - *Definition*.- Let $\phi : G \times N \to N$ be a continuous action of the group G on the manifold N. A foliation (N,F) is <u>invariant under</u> ϕ if $\phi : G \times N^\delta \to N^\delta$ is continuous (that is each element of G acts as an F-isomorphism).

If $\pi : M' \to M$ is a regular covering with group G and F is a foliation on M then the induced foliation π^*F on M' is of course invariant under the action of G on M' as group of covering translations of π. This yields the desired condition.

2.2.5. - *Lemma*. - Let $\pi : M' \to M$ be a regular covering with group of covering translations G and let F' be a foliation on M' invariant under the action of G. Then there is a unique foliation F on M such that $F' = \pi^* F$.

The lemma is proved in just the same way as lemma I; 3.1.3.

2.2.6. - *Corollary*.- Let $M \xleftarrow{\pi} \tilde{M} \xrightarrow{D} F^n$ be an equivariant submersion and let F^* be the foliation on $\tilde{M}$ given by the submersion D. Then there exists a unique foliation F of codimension n on M such that $\pi^* F = F^*$.

We conclude this section with two important applications of the concept of transversality, the first one of which is a theorem ascribed to Ehresmann [Eh].

Given a foliated bundle $\xi = (M,p,B)$ there are two transverse foliations on M,

(1) the fibration itself and

(2) the foliation defined in 2.1.7, ii).

It is natural to ask whether any fibre bundle $\xi = (M,p,B)$ with a foliation transverse to the fibres arises in this way.

2.2.7. - *Theorem*.- Let $\xi = (M,p,B)$ be a fibre bundle with fibre F and let F be a foliation on M transverse to the fibration. If F is compact then ξ is a foliated bundle whose (transverse) foliation is given by F.

Proof : Let $b \in B$ and F_b the fibre over b. As F_b is compact we may cover it by a finite number of open cubes $U_1,\dots,U_s$ which are bidistinguished with respect to F and the fibration.

Put $V = \bigcap_{i=1}^{s} p(U_i)$. As p is an open map it follows that

V is open and the preimage $p^{-1}(V)$ is contained in $\bigcup_{i=1}^{s} U_i$. If F_V is the foliation on $p^{-1}(V)$ which is induced by F then it is not hard to see that each leaf of F_V intersects each fibre F_b, $b \in V$, in exactly one point. Hence there is defined a local trivialization

$$\phi_V : p^{-1}(V) \longrightarrow V \times F$$

such that F_V is given by the submersion

$$f_V = pr_F \circ \phi_V : p^{-1}(V) \to F .$$

This shows that the cocycle corresponding to the fibre bundle atlas $\{(V,\phi_V)\}$ is locally constant. □

See exercise 2.2.9, ii) to affirm that the compactness of F in the last theorem is a necessary condition.

The second application of transversality concerns orientability. If (as in the case $m = 2$) T_o denotes the topology on $\mathbb{R}^m = \mathbb{R}^{m-n} \times \mathbb{R}^n$ which makes all distinguished charts of a foliation (M,F) homeomorphisms with respect to the leaf topology on M, i.e. $(\mathbb{R}^m,T_o) = \mathbb{R}^{m-n} \times (\mathbb{R}^n)^\delta$, then $(\mathbb{R}^m,T_o)$ is a canonically oriented (non-countable) $(m-n)$-dimensional manifold.

2.2.8. - *Definition.* - Let (M,F) be a foliation of codimension n.

i) We say that F is transversely orientable if it is represented by a foliated atlas whose corresponding foliated cocycle has values in H^o_{n+}, the pseudogroup of orientation preserving local homeomorphisms of $\mathbb{R}^n$.

ii) The foliation F is orientable if F is represented by a foliated atlas $A = \{(U_i,\varphi_i)\}$ such that for all pairs (i,j) the local homeomorphism $\varphi_i \circ \varphi_j^{-1} : (\mathbb{R}^m,T_o) \to (\mathbb{R}^m,T_o)$ is orientation preserving.

A foliation F being orientable intuitively means that the leaves of F can be oriented in a coherent manner (cf. also 2.3.2.). Note that the fact that all leaves of a foliation F are orientable in general does not imply that F is orientable.

The definitions I; 2.3.4. of the <u>transverse orientation covering</u> $\pi : M^* \to M$ and I; 2.3.5. of the <u>tangent orientation covering</u> $\tau : M_* \to M$ translate verbatim to arbitrary foliated manifolds (M,F). Proposition I; 2.3.6, including its proof, generalizes immediately, i.e.

i) π^*F is always transversely orientable, and F is transversely orientable if and only if π is trivial.

ii) τ^*F is always orientable, and F is orientable if and only if τ is trivial.

2.2.9. - *Exercises*. - i) Give an example of a foliation which does not admit any transverse foliation of complementary dimension.

ii) On $\mathbb{R}^2$ consider the product fibration $(\mathbb{R}^2, pr, \mathbb{R})$, where pr is the projection onto the x-axis, and a foliation F, as indicated in figure 3.

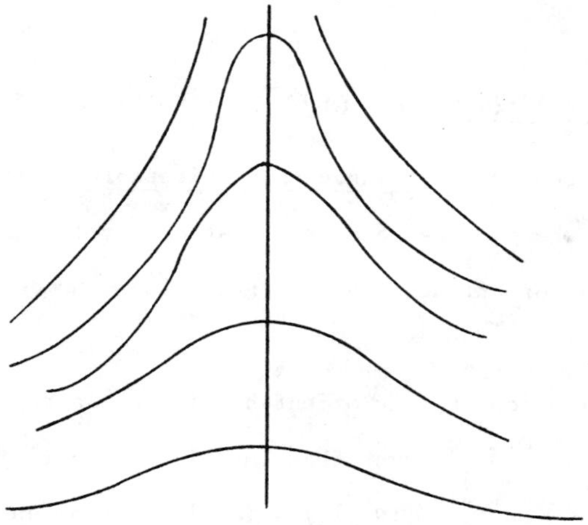

Figure 3

Find a submersion that defines a foliation of this type and show that $\mathbb{R}^2$ endowed with this foliation is not a foliated bundle.

iii) Let (M,F) be a compact foliated manifold of codimension one. A <u>closed transversal</u> of F is a submanifold θ homeomorphic to S^1 and transverse to F.

Show that for every $x \in M$ such that L_x is non-compact there exists a closed transversal of F passing through x.

iv) Show that the Reeb foliations on S^3 and $S^1 \times S^{m-1}$ are both orientable and transversely orientable. The compact leaf does not admit any closed transversal.

v) (Cf. I; 2.3.10). Let (M,F) be a foliated manifold. If there exists an orientable foliation $F^{\pitchfork}$ transverse to F then F is transversely orientable.

2.3. *The tangent bundle of a foliation ; Frobenius' theorem.*

In this and the next section the reader is supposed to have some elementary knowledge of vector bundles and differential forms on manifolds. For reference on this subject see Milnor - Stasheff [MS] and Sternberg [Ste], for example.

Everything that follows holds in any differentiability class C^r, $r \geq 2$.

Given a foliation (M,F) we want to associate to it a subbundle of the tangent bundle (TM,p,M) of M, the so-called tangent bundle of F.

2.3.1. - *Definitions and remarks.* - Let (M,F) be a foliated manifold of codimension n .

i) A vector $v \in TM$ is <u>tangent</u> to the leaf $L_x \in F$ through

$x = p(v)$ if there exists a distinguished map (U,f) of F around x such that $T_x f(v) = 0$.

It is not hard to see that the set of vectors tangent to the leaves of F forms the total space of a subbundle TF of TM. It is called the <u>tangent bundle of</u> F.

ii) The <u>normal bundle</u> of F, denoted NF, is by definition the normal bundle of TF (, i.e. the cokernel of $TF \hookrightarrow TM$).

Certain properties of foliations can be read off from TF and NF, for example :

2.3.2. - <u>*Exercises*</u>. - i) An ℓ-dimensional foliation F of codimension n is orientable (resp. transversely orientable) if and only if TF (resp. NF) is orientable (, i.e. its structure group $GL(\ell;\mathbb{R})$ resp. $GL(n;\mathbb{R})$ can be reduced to the subgroup consisting of the elements with positive determinant).

ii) (Cf. I; 2.3.11). Let F be an orientable foliation on M. Then M is orientable if and only if F is transversely orientable.

As we have seen, to every foliated manifold (M,F) we can assign a subbundle TF of TM. It is therefore natural to ask whether every subbundle of TM is the tangent bundle of a foliation on M. The answer to this question is provided by "Frobenius' theorem" (which, as Milnor has pointed out, is due not to Frobenius, but was known already to Clebsch and Deahna). It is derived next.

2.3.3. - <u>*Definition*</u>. - Let M be an m-dimensional manifold and $\xi = (E,p',M)$ an <u>ℓ-plane bundle</u>, i.e. a subbundle of rank ℓ of TM.

i) A vector field X on the open set $U \subset M$ is <u>tangent</u> to ξ if $X(x) \in E$, for every $x \in U$.

ii) The ℓ-plane bundle ξ is called <u>involutive</u> if for any $x \in M$ there is a neighbourhood U of x in M such that $[X,Y]$ is tangent to ξ for any two vector fields X, Y defined on U and tangent to ξ.

iii) We denote by E_x the fibre of ξ over the point $x \in M$. A submanifold P of M is called an <u>integral manifold</u> of ξ if the inclusion $i_P : P \to M$ satisfies the condition

$$T_x i_P(T_x P) = E_x,$$

for every $x \in P$.

iv) ξ is <u>(completely) integrable</u> if for every $x \in M$ there is a chart

$$\varphi : U \to \mathbb{R}^m = \mathbb{R}^\ell \times \mathbb{R}^n$$

of M, a so-called <u>distinguished chart of</u> ξ, such that for each $y \in \mathbb{R}^n$ the submanifold $\varphi^{-1}(\mathbb{R}^\ell \times \{y\})$ of M is an integral manifold (a <u>plaque</u>) of ξ.

2.3.4. - <u>*Lemma*</u>. - <u>A subbundle ξ of TM is integrable if and only if it is the tangent bundle of a foliation on</u> M.

<u>*Proof*</u> : Let ξ be an integrable subbundle of TM and (U,φ) a distinguished chart of ξ. A rank argument shows that $T_\varphi(\xi|U)$ is the tangent bundle of the horizontal foliation on $\mathbb{R}^\ell \times \mathbb{R}^n$ (leaves $\mathbb{R}^\ell \times \{y\}$) and thus for $f = pr_{mn} \circ \varphi$ we get $Tf(\xi|U) = 0$. This implies that, for each connected integral manifold P of $\xi|U$, we have $f(P) = \text{const}$.

Now let (U_i,φ_i) and (U_j,φ_j) be two distinguished charts of ξ with $U_i \cap U_j \neq \emptyset$ and let Q be a plaque in U_i. If R is a component of $Q \cap U_j$ then, by what was just proved, R is contained in a plaque of U_j. This means that there exists a local homeomorphism $\gamma_{ij} \in H_n^1$ such that $\gamma_{ij} \circ f_j = f_i$ and thus there is defined a foliated cocycle.

This proves the lemma, for the tangent bundle of a foliation evidently is integrable. □

2.3.5. - *Frobenius' Theorem*. - An ℓ-plane bundle $\xi \subset TM$ is integrable (hence the tangent bundle of a foliation) if and only if it is involutive.

Proof : Using a foliated atlas it is not hard to see that an integrable subbundle of TM is involutive. To show the converse we proceed by two steps.

a) Let $\xi \subset TM$ be an ℓ-plane bundle and let $Y_1,\ldots,Y_\ell$ be ℓ linearly independent vector fields spanning $\xi|U$ over the sufficiently small chart neighbourhood U. In local coordinates, Y_λ is given by

$$Y_\lambda = \sum_{\mu=1}^{m} a_{\lambda\mu} \frac{\partial}{\partial x_\mu}, \quad 1 \leq \lambda \leq \ell,$$

with differentiable maps $a_{\lambda\mu}$ on U.

As the Y_λ are linearly independent the matrix $(a_{\lambda\mu})$ has rank ℓ. Thus (possibly after reindexing the $\frac{\partial}{\partial x_\mu}$ and making U smaller) we may suppose that

$$A(x) = (a_{\lambda\mu}(x)), \quad 1 \leq \lambda,\mu \leq \ell,$$

is invertible for all $x \in U$. Let

$$B(x) = (b_{\lambda\mu}(x)) = A^{-1}(x) .$$

Then the maps $b_{\lambda\mu}$ are differentiable on U, $1 \leq \lambda,\mu \leq \ell$. Set

$$X_\lambda = \sum_{\mu=1}^{\ell} b_{\lambda\mu} Y_\mu,$$

then X_λ can be written as

$$(1) \qquad X_\lambda = \frac{\partial}{\partial x_\lambda} + \sum_{\mu=\ell+1}^{m} c_{\lambda\mu} \frac{\partial}{\partial x_\mu},$$

with differentiable maps $c_{\lambda\mu}$ on U. Moreover, the X_λ are linearly independent on U. As ξ is involutive we get

$$(2) \qquad [X_\lambda, X_\mu] = \sum_{i=1}^{\ell} d_i X_i,$$

with differentiable maps d_i on U.

Since $\left[\frac{\partial}{\partial x_\lambda}, \frac{\partial}{\partial x_\mu}\right] = 0$ it follows from (1) and (2) that $[X_\lambda, X_\mu] = 0$, $1 \leq \lambda$, $\mu \leq \ell$.

b) For $\varepsilon > 0$, let D_ε be the open ε-disk in $\mathbb{R}^n$, $\ell + n = m$, let $x_o \in U$ and let $h : D_\varepsilon \to M$ be an embedding, $h(0) = x_o$, and such that h is transverse to ξ, i.e. for each $y \in D_\varepsilon$ the tangent space of $h(D_\varepsilon)$ in $x = h(y)$ and ξ_x span $T_x M$.

We denote by φ_t^λ the local flow of X_λ, $1 \leq \lambda \leq \ell$. As $[X_\lambda, X_\mu] = 0$ we have $\varphi_t^\lambda \circ \varphi_s^\mu = \varphi_s^\mu \circ \varphi_t^\lambda$ for all λ, μ. There is a neighbourhood V of 0 in $\mathbb{R}^\ell$ and a well defined map

$$\psi : V \times D_\varepsilon \to M$$

$$(t_1, \ldots, t_\ell, y) \mapsto \varphi_{t_1}^1 \circ \ldots \circ \varphi_{t_\ell}^\ell (h(y)).$$

As $T_o\psi(\frac{\partial}{\partial t_\lambda})_o = X_\lambda(x_o)$ and h is transverse to ξ we conclude that ψ has rank m in 0. Thus ψ is a local diffeomorphism in an open neighbourhood of 0. Then for a small open neighbourhood U' of x_o (U', ψ^{-1}) is a distinguished neighbourhood of ξ. □

2.3.6. - Remark. - In the proof of Frobenius' theorem we have essentially used that the vector fields under consideration are at least C^2. The theorem holds however also for C^1 plane fields, cf. say Camacho-Neto [CN].

2.4. *Pfaffian forms ; Frobenius' theorem (dual version).*

We now give another version of Frobenius' theorem in terms of differentiable 1-forms or Pfaffian forms.

For this let M be an m-manifold of class C^r, $r \geq 2$, and TM its tangent bundle. A Pfaffian form on M is a differentiable map

$$\omega : TM \to \mathbb{R}$$

which restricted to each fibre T_xM is linear. It is without singularities if its restriction to each fibre is not identically zero.

An n-tuple $\Omega = (\omega^1,\ldots,\omega^n)$ of Pfaffian forms is a Pfaffian system of rank n on M if the map

$$\Omega : TM \to \mathbb{R}^n$$

has rank n when restricted to any fibre of TM.

If we consider $\mathbb{R}^n$ as the trivial vector bundle over a point as base then Ω is a vector bundle homomorphism.‡ Thus $\ker \Omega$ is a subbundle of TM which is called the kernel of the system Ω.

2.4.1. - Lemma. - A subbundle ξ of TM is the kernel of a Pfaffian system if and only if the normal bundle $N\xi$ is trivial.

Proof : This is obvious, for $N\xi$ is the cokernel of $\xi \hookrightarrow TM$. □

---

‡ In contrast to fibre bundle maps, vector bundle homomorphisms need not be homeomorphisms when restricted to fibres.

The preceding lemma shows that a plane bundle $\xi \subset TM$ is locally, though not necessarily globally, the kernel of a Pfaffian system, because ξ and $N\xi$ are locally trivial.

It remains to see how the condition of involutiveness translates into the language of Pfaffian forms. For this we use the following elementary result. (The manifold V may be thought of as an open subset of M .)

2.4.2. - *Lemma*. - Let $\Omega = (\omega^1,\dots,\omega^n)$ be a Pfaffian system of rank n on the m-manifold V. If η is an arbitrary 2-form on V then the following two conditions are equivalent.

(1) $\eta \wedge \omega^1 \wedge \dots \wedge \omega^n = 0$,

(2) there are 1-forms $\alpha^1,\dots,\alpha^n$ on V such that $\eta = \sum_{i=1}^{n} \alpha^i \wedge \omega^i$.

Proof : Evidently, (2) implies (1).

Suppose $\Omega = (\omega^1,\dots,\omega^n)$ is a Pfaffian system of rank n on V. Locally, we can complete Ω to a trivialization of the cotangent bundle T^*V of V. Let $\bar{\Omega}_U = (\omega^1,\dots,\omega^n,\omega^{n+1},\dots,\omega^m)$ be such a completion over $U \subset V$. Then

$$\eta|U = \sum_{i<j\leq m} a_{ij}\omega^i \wedge \omega^j .$$

But condition (1) implies $a_{ij} = 0$ for $i > n$. It suffices therefore to put

$$\alpha^i = - \sum_{i<j} a_{ij}\, \omega^j$$

and (2) holds on U. Using a partition of unity we get the coefficients a_{ij} globally on V. □

2.4.3. - *Proposition*. - Let $\xi \subset TV$ be an ℓ-plane bundle which is the kernel of the Pfaffian system $\Omega = (\omega^1,\dots,\omega^n)$. Then ξ is integrable if and only if the following condition is satisfied:

(*) $$d\omega^i \wedge \omega^1 \wedge \ldots \wedge \omega^n = 0 \quad \underline{\text{for each}}\ i, \quad 1 \leq i \leq n,$$

Proof : Suppose condition (*) holds. By the preceding lemma we have

$$d\omega^i = \sum_{j=1}^{n} \alpha_i^j \wedge \omega^j.$$

Thus if X and Y are two vector fields tangent to ξ then

$$d\omega^i(X,Y) = 0 \quad \text{for each} \quad i.$$

This implies

$$\omega^i([X,Y]) = 0 \quad \text{for each } i,$$

and hence ξ is involutive.

Conversely, for $x \in V$ we take linearly independent vector fields $X_1,\ldots,X_m$ which are defined in a neighbourhood U of x in V and such that $X_1,\ldots,X_\ell$ span $\xi|U$. Then, by the involutiveness of ξ it is easily seen that the left hand side of (*) evaluated on (n+2)-tupels of the X_i vanishes. This proves the proposition. □

If $\xi \subset TM$ is a plane bundle and $U \subset M$ is a trivializing open set then $\xi|U$ is the kernel of a Pfaffian system $(\omega^1,\ldots,\omega^n)$ on U.

The dual version of the Frobenius theorem is now a consequence of 2.4.3.

2.4.4. - *Theorem*. - The ℓ-plane bundle $\xi \subset TM$ is integrable if and only if every $x \in M$ has a neighbourhood U on which there exists a Pfaffian system $\Omega_U = (\omega^1,\ldots,\omega^n)$ whose kernel is $\xi|U$ and such that

$$d\omega^i \wedge \omega^1 \wedge \ldots \wedge \omega^n = 0 \quad \underline{\text{for any}}\ i, \ 1 \leq i \leq n.$$

In particular, in the case of codimension one lemma 2.4.1 implies that the following conditions are equivalent.

(1) $N\xi$ is trivial.

(2) there exists a Pfaffian form ω on M such that $\xi = \ker \omega$.

(3) $N\xi$ is orientable.

Such an $(m-1)$-plane bundle is the tangent bundle of a foliation if and only if $d\omega \wedge \omega = 0$ or $d\omega = \alpha \wedge \omega$ for some 1-form α on M.

Exercise.- Let $\omega = z dx + x dy + y dz$ be a 1-form on $\mathbb{R}^3$.

i) Determine the submanifold M of $\mathbb{R}^3$ where ω does not vanish.

ii) Is ω integrable on M (i.e. $d\omega \wedge \omega = 0$) ?

3. *Examples of foliated manifolds.*

In this paragraph, we shall describe two interesting classes of foliations. Both are related to the examples of §1 in so far as they can be defined by global data. Some familiarity with the elements of Lie group theory will help the reader.

3.1. *Foliations defined by locally free group actions.*

In what follows everything is assumed to be of class C^1.

Let

$$\Phi : G \times M \to M$$

be an action of the (connected) Lie group G on the m-manifold M. We write simply $g(x)$ for $\Phi(g,x)$. For $x \in M$, we denote by $G(x)$ the orbit of x under G.

Replacing G by its universal covering, we may suppose that G is simply connected. This will be no loss of generality for our purposes.

3.1.1. - *Definition*. - The action $\Phi : G \times M \to M$ is called

i) locally free if, for every $x \in M$, the isotropy group G_x of G in x is discrete,

ii) free if $G_x = \{e\}$ for each $x \in M$.

Now let $\Phi : G \times M \to M$ be a locally free action of the ℓ-dimensional Lie group G on M.

i) If Φ_x is the restriction of Φ to $G \times \{x\}$ one has the commutative diagram

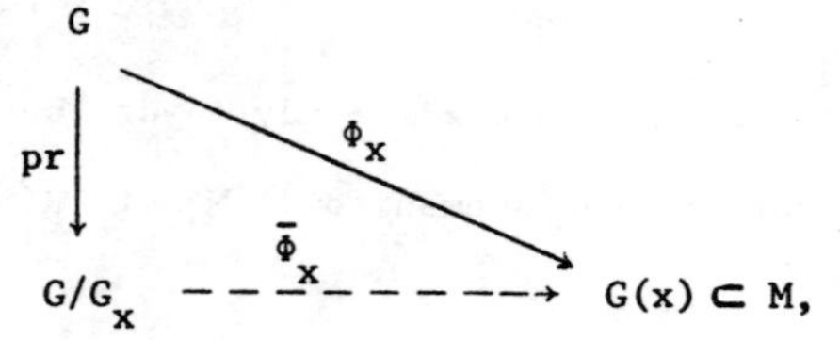

where pr is the projection and the induced map $\bar{\Phi}_x$ is an injective C^r immersion. In other words, the orbits of G are injectively immersed C^r submanifolds of M and diffeomorphic to (right) homogeneous spaces of G.

ii) Furthermore, if $x' \in G(x)$, $x' = g(x)$, then

$$G_x = g^{-1} G_{x'} g$$

and one has the commutative diagram

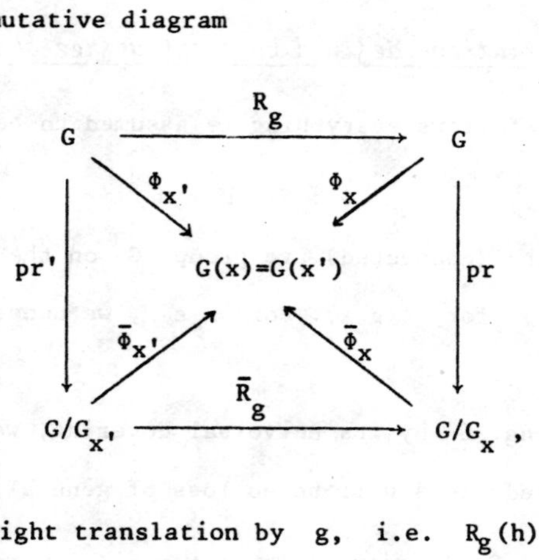

where R_g denotes right translation by g, i.e. $R_g(h) = hg$, and $\bar{R}_g$ is canonically induced by R_g. This shows that the structure of $G(x)$ as (right) homogeneous space of G does not depend on the point x.

Next we introduce the Lie algebra $\mathcal{g}$ of <u>right invariant vector fields</u> on G, i.e. $X \in \mathcal{g}$ if and only if $TR_g \circ X = X \circ R_g$, for each $g \in G$. For $X \in \mathcal{g}$, the vector field $(X,0)$ on $G \times M$ again is denoted by X.

iii) For each $(g,x) \in G \times M$, the vector $T\Phi(X(g,x)) \in TM$ is tangent to the orbit $G(x)$ at $g(x)$ and, by the remarks i) and ii) above, if $g(x) = g'(x')$ one has

$$T\Phi(X(g,x)) = T\Phi(X(g',x')).$$

Hence there is defined a <u>complete</u> (i.e. with global 1-parameter group) vector field $\bar{X}$ on M such that $\bar{X}(x)$ is tangent to $G(x)$, for each $x \in M$.

iv) Moreover, if $\mathcal{X}(M)$ denotes the algebra of vector fields on M

$$\psi : \mathcal{g} \to \mathcal{X}(M),$$

defined by $\psi(X) = \bar{X}$, is a homomorphism of Lie algebras. (It is called the homomorphism <u>associated</u> to the action Φ.)

v) Let $\{X_1,\ldots,X_\ell\}$ be a basis of $\mathcal{g}$. As the action Φ is locally free the set of vector fields $\{\bar{X}_1,\ldots,\bar{X}_\ell\}$ is of rank ℓ in each point of M. It defines an ℓ-plane bundle ξ over M which is trivial since it is globally defined by linearly independent vector fields and which is moreover integrable, by iv) and Frobenius'theorem, see 2.3.5. and 2.3.6.

vi) Let F be the foliation of dimension ℓ on M such that $\xi = TF$. As $\dim G = \ell$ it follows from iii) that the leaf L_x through the point $x \in M$ coincides with $G(x)$.

3.1.2. - *Definition*. - The foliation F with $TF = \xi$ is called <u>the foliation defined by the (locally free) action</u> Φ <u>of</u> G <u>on</u> M.

The existence of the foliation $\mathcal{F}$ was derived from the properties of the homomorphism of Lie algebras

$$\psi : \mathfrak{g} \to \mathcal{X}(M).$$

It is hence interesting to know whether conversely each such homomorphism defines an action of the simply connected Lie group G on M. We have :

3.1.3. - *Proposition*. - Let

$$\psi : \mathfrak{g} \to \mathcal{X}(M)$$
$$X \mapsto \bar{X}$$

be a homomorphism of Lie algebras such that

(1) each element Y of $\operatorname{im}\psi$ is complete,

(2) $\operatorname{im}\psi$ has rank ℓ in every point of M, where $\ell = \dim G$.

Then there is a unique locally free action Φ of the simply connected Lie group G (with Lie algebra $\mathfrak{g}$) on M such that ψ is the homomorphism associated to Φ.

Proof : We consider the subbundle $\tilde{\xi}$ of $T(G \times M) = T(G) \times T(M)$ which is generated by the set of pairs of vector fields $(X,\bar{X})$, with $X \in \mathfrak{g}$. This is an ℓ-plane bundle which is moreover integrable, by the Frobenius theorem, and thus yields a foliation $\tilde{\mathcal{F}}$ on $G \times M$. As the vector fields X and $\bar{X}$ are complete it follows that the restriction of pr_G to any leaf $\tilde{L} \in \tilde{\mathcal{F}}$ is a covering map onto G and thus a diffeomorphism, since G is simply connected.

The foliation $\tilde{\mathcal{F}}$ yields the free action

$$\tilde{\Phi} : G \times (G \times M) \to G \times M$$
$$(g,(h,x)) \mapsto \tilde{L}_{(h,x)} \cap (\{gh\} \times M) ,$$

(that is $\tilde{\Phi}$ is defined by lifting equivariantly the group structure of G to each leaf $\tilde{L}$ of $\tilde{F}$ by means of pr_G).

The action $\tilde{\Phi}$ preserves the product structure of $G \times M$. Indeed, $\tilde{\Phi}$ preserves the vertical foliation, by definition. On the other hand, $\tilde{F}$ is preserved by right translation on G, i.e. by

$$\begin{aligned} R : G \times (G \times M) &\to G \times M \\ (g,(h,x)) &\mapsto (hg,x) . \end{aligned}$$

Moreover, for points (h,x) and (k,x) of $G \times M$ we have

$$\tilde{L}_{(k,x)} = R_{(h^{-1}k)}(\tilde{L}_{(h,x)})$$

and therefore

$$\tilde{L}_{(k,x)} \cap (\{gk\} \times M) = R_{(h^{-1}k)}(\tilde{L}_{(h,x)} \cap (\{gh\} \times M).$$

Hence $\tilde{\Phi}$ preserves the horizontal foliation on $G \times M$, also, i.e. the second component of $\tilde{\Phi}(g,(h,x))$ is independent of h. This defines the desired action Φ of G on M. As $\operatorname{im} \phi$ is of maximal rank it follows that Φ is locally free.

The associated homomorphism

$$\tilde{\phi} : \mathfrak{g} \to \mathcal{X}(G \times M)$$

of $\tilde{\Phi}$ is defined by $\tilde{\phi}(X) = (X,\bar{X})$. Indeed, $\tilde{\Phi}$ is the lift to $\tilde{F}$ of the left translation

$$\begin{aligned} L : G \times (G \times M) &\to G \times M \\ (g,(h,x)) &\mapsto (gh,x) . \end{aligned}$$

Therefore $\tilde{\phi}(X)$ is the lift of X which is $(X,\bar{X})$, and $\phi(X) = \bar{X}$ is the associated homomorphism of Φ as required. □

We conclude this section with some examples, remarks and exercises.

3.1.4. - *Examples*.- i) Every complete vector field without singularities on M defines a locally free action of $\mathbb{R}$ on M.

ii) Let H be a Lie group, G a closed subgroup of H and Γ a discrete subgroup of H. Then G acts by left translation on H/Γ. This action is of course locally free.

3.1.5. - *Remarks*.- The previous theory can be developed for local actions of G on M, i.e. defined only in neighbourhoods of the points of M. As in the proof, part b) of 2.3.5., we have only used that ℓ commuting vector fields define a local action of $\mathbb{R}^\ell$ and that such an action, if it is free, determines a foliation on M.

3.1.6. - *Exercises*.- i) Let $\Phi : G \times M \to M$ be a free action of a compact Lie group. Then the foliation $\mathcal{F}$ defined by Φ is in fact a G-principal bundle over a certain manifold B.

ii) Construct a locally free action of $\mathbb{R}^2$ on the 3-torus T^3 such that all leaves of the corresponding foliation are dense in T^3.

iii) Let $H : \pi_1 T^2 \to \mathrm{Diff}_+(S^1)$ be a representation and ξ_H the corresponding suspension. Show that the foliation of ξ_H is defined by a locally free action of $\mathbb{R}^2$.

iv) Show that the Reeb component on $D^2 \times S^1$ cannot be defined by a locally free action of any Lie group.

3.2. *Foliations with a transverse structure.*

With our last family of examples of foliations we come back to the class of equivariant submersions studied in section 1.4.

For this we consider a topological group G acting effectively on the manifold F.

3.2.1. - *Definition*. - Let M be a manifold. A pair $C = (\{(U_i,f_i),\{g_{ij}\})$ is a *(foliated) cocycle on* M *with values in* (G,F) if C satisfies definition 2.1.4. with $\mathbb{R}^n$ replaced by F and $\mathcal{H}_n^r$ replaced by G.

A cocycle C with values in (G,F) certainly defines a foliated cocycle in the sense of 2.1.4. We only need to refine C so that the sets $f_i(U_i)$ come to lie in coordinate neighbourhoods of F. Thus C determines a foliation of codimension n on M, where n is the dimension of F.

We now want to describe these foliations by means of global data. First an example.

Example. Let $\mathcal{F}$ be a foliation on M defined by a surjective equivariant submersion

$$\begin{array}{ccc} \tilde{M} & \xrightarrow{\ D\ } & F \\ {\scriptstyle\pi}\big\downarrow & & \\ M & & \end{array}$$

If the group of covering translations Γ of π acts effectively on F then it is easy to see that $\mathcal{F}$ can be defined by a cocycle with values in (Γ,F); see 1.4.3.

We are interested in the converse situation.

3.2.2. - *Theorem*. - *Every foliation* $\mathcal{F}$ *on* M *defined by a cocycle with values in* (G,F) *is defined by a surjective equivariant submersion* $F \longleftarrow \tilde{M} \longrightarrow M$.

Proof :

a) Suppose $\mathcal{F}$ is given by the cocycle $C = (\{U_i, f_i)\}, \{g_{ij}\})$. We consider the cocycle $\hat{C} = (\{U_i\}, \{g_{ij}\})$ with values in G (in the sense of 1.1.). The G-bundle $\xi_{\hat{C}} = (E,p,M)$ with fibre F, as constructed in 1.1.11., is a foliated bundle, for $\{g_{ij}\}$ is locally constant. We denote by $\mathcal{F}_E$ the corresponding foliation on E.

For each i, let

$$\begin{aligned} \sigma_i : U_i &\to U_i \times F \\ x &\mapsto (x, f_i(x)) \end{aligned}$$

be a local section. Recalling the construction of $\xi_{\hat{C}}$ one sees that the maps σ_i fit together to define a section

$$\sigma : M \to E$$

which is transverse to $\mathcal{F}_E$ and such that $\sigma^*(\mathcal{F}_E) = \mathcal{F}$.

b) Now let

$$H : \pi_1 M \to G$$

be the holonomy of the foliated bundle $\xi_{\hat{C}}$ and let $\pi : \tilde{M} \to M$ be the covering map belonging to ker H. The induced bundle $\pi^* \xi_{\hat{C}}$ is the trivial bundle and we get the following commutative diagram

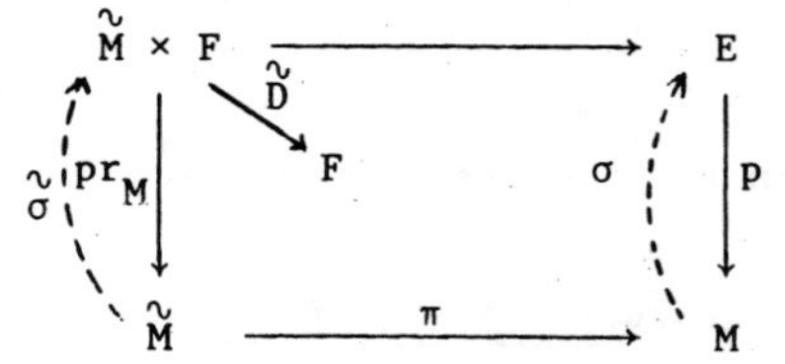

where the section $\tilde{\sigma}$ is constructed as the section σ in a) and $\tilde{D}$ is the canonical projection. This proves the theorem. □

3.2.3. - *Examples*. i) If G is a Lie group and $F = G/K$, where K is a closed subgroup, then the corresponding foliations are called transversely homogeneous foliations.

For example, if $\mathcal{F}$ is a foliation defined by an action of a Lie group and if $\mathcal{F}'$ is transverse to $\mathcal{F}$ then $\mathcal{F}'$ is transversely homogeneous. (This is the case for every vector field transverse to a foliation of codimension one defined by a Lie group action).

The class of transversely homogeneous foliations has been studied by Blumenthal [Bl].

ii) If G is a Lie group and $F = G$ then the corresponding foliations are called transversely Lie foliations. The easiest examples of this type are the foliations of codimension one that are defined by a closed 1-form. (These are transversely Lie with group $\mathbb{R}$). We shall study these foliations in chapter...

For more detailed information about transversely Lie foliations we refer the reader to the articles by Hermann [Her] and Fedida [Fe].

We content ourselves with two remarks concerning these foliations.

3.2.4. - *Definition* (Reinhart [Rei]). - Let μ be a riemannian metric on the foliated manifold $(M,\mathcal{F})$ of class C^r, $r \geq 1$, and codimension n. We think of $N\mathcal{F}$ as the orthogonal complement of $T\mathcal{F}$. The metric μ is bundle-like if there exists a riemannian metric on $\mathbb{R}^n$ and $\mathcal{F}$ is given by a foliated cocycle $(\{(U_i,f_i)\},\{g_{ij}\})$ such that

(1) $Tf_i : N\mathcal{F}|U_i \to T\mathbb{R}^n$ is an isometry,

(2) the maps $g_{ij}(x)$, $x \in M$, are local isometries of $\mathbb{R}^n$.

Intuitively, a metric μ on (M,F) is bundle-like if the "distance" between two plaques taken along orthogonal transversals is constant.

The foliations with a bundle-like metric are studied in Reinhart [Rei]. We use this notion to prove a final result.

3.2.5. - *Proposition*. - Let F be a transversely Lie foliation on a compact manifold M. Then F is defined by an equivariant (locally trivial) fibration.

Sketch of proof : a) Given a riemannian metric on M, one chooses a left invariant metric on the group G, lifts it by means of the distinguished maps f_i to the orthogonal complement of TF and so constructs a bundle-like metric μ on (M,F). As M is compact it follows that μ is complete.

b) On the other hand, by the preceding theorem, one knows that F is defined by an equivariant submersion $(M,F) \xleftarrow{\pi} \tilde{M} \xrightarrow{D} G$. The foliation $\tilde{F} = \pi^* F$ supports a bundle-like metric, namely $\tilde{\mu} = \pi^* \mu$. Using the fact that μ is complete one shows that D is locally trivial. □

3.2.6. - *Remark*. - The above proposition leads to a new definition of transversely Lie foliations. In fact, let α be the Maurer-Cartan form on G ; it satisfies the equation

$$d\alpha + \frac{1}{2}[\alpha,\alpha] = 0.$$

Then $\omega = D^*\alpha$ is a differential 1-form on $\tilde{M}$ with values in the Lie algebra $\mathfrak{g}$ of G and of maximal rank in each point. The form ω has the following properties :

(1) $d\omega + \frac{1}{2}[\omega,\omega] = 0$,

(2) ω defines $(\tilde{M},\tilde{F})$ in the sense that $T\tilde{F} = \ker \omega$.

(3) ω is invariant under the group of covering translations of $(\tilde{M},\pi,M)$.

By condition (3), ω induces a 1-form Ω on M with values in $\mathfrak{g}$ that satisfies the Maurer-Cartan equation and defines the foliation F.

Conversely, by the Lie theorem, such a form Ω on M yields a cocycle with values in (G,G).

3.2.7. - *Exercise*.- Let F be a foliation of class C^2 on T^2. Show that the following conditions are equivalent :

(1) F is defined by a closed 1-form.

(2) F is transversely Lie,

(3) F admits a bundle-like metric.

Is the same **true** for a codimension one foliation on T^m, $m \geq 3$?

CHAPTER III

HOLONOMY

The concept of holonomy is of fundamental interest in the theory of foliations. Its introduction by Ehresmann (see Haefliger [Ha; p.377]) may be understood in some way as the beginning of foliation theory as a distinct field of research.

In this chapter we define the notion of holonomy for arbitrary foliations. This notion already occured in the first two chapters but with apparently different meanings.

In chapter I we studied the holonomy of circle leaves on surfaces. It turned out that the holonomy of such a leaf L determines the foliation in a neighbourhood of L; see I; 3.2. We also observed that holonomy could be introduced for leaves other than circles but would always be trivial there.

Chapter II to a large extent dealt with the study of the holonomy representations of foliated bundles. We saw that foliated bundles are completely determined by their holonomy.

In general, for an arbitrary foliated manifold $(M,\mathcal{F})$ holonomy cannot be defined globally for all of $\mathcal{F}$ as a representation of the fundamental group of M in some group, but only for the leaves individually. Also simple examples show that it is only for leaves L of $\mathcal{F}$ belonging to a special class that the holonomy of L determines the foliation in a neighbourhood of L; see 2.1.7.

In order to clarify the relation between the two different notions of holonomy discussed in chapters I and II and in order to show how they fit into the general framework we shall associate (in 2.1) to each leaf L of a foliation (M,F) a so-called regularly foliated microbundle μ_L and then define the holonomy of L in terms of μ_L. Roughly, a regularly foliated microbundle of rank n over a space L is an "arbitrarily small" neighbourhood E of the zero-section in an $\mathbb{R}^n$- bundle over L, together with a foliation F_L of codimension n on E which is transverse to the fibres and contains the zero-section as a leaf, cf. 1.2.1, 1.2.7.

In the differentiable case, we can give a more precise idea of what we have in mind. As $\mathbb{R}^n$- bundle we take the normal bundle ν_L of L. A sufficiently small neighbourhood E of the zero-section L of ν_L is then mapped by the exponential map α locally diffeomorphically onto some open neighbourhood of L in M and may be endowed with the foliation $F_L = \alpha^* F$. As there is no canonical choice for E , we pass to the germ of (E,F_L) near L. This gives us the foliated microbundle μ_L. The advantage of this procedure is that now L has become "unwrapped" in the sense that it is a leaf of F_L which is closed in E.

We work in a fixed differentiability class C^r , $r \geq 0$, unless otherwise stated.

1. *Foliated microbundles*.

Before introducing the general notion of foliated microbundles and their holonomy representations (in 1.2 and 1.3) we want to illustrate in 1.1 our strategy by describing the holonomy of leaves of foliated bundles in such a way that it becomes a special case of the general concept. For that purpose we first construct a regularly foliated microbundle μ_L for leaves of foliated bundles and show which properties of foliated bundles still hold for μ_L .

1.1. *Localization in foliated bundles.*

1.1.1. *Pseudobundles and microbundles associated to leaves in foliated bundles.*

i) Let $\xi = (M,p,B)$ be a foliated bundle with n - dimensional fibre F and transverse foliation $\mathcal{F}$. When L is a leaf of $\mathcal{F}$ we have the induced diagram

$$\begin{array}{ccc} \hat{M} & \xrightarrow{\ f\ } & M \\ {\scriptstyle \hat{p}}\downarrow & & \downarrow{\scriptstyle p} \\ L & \xrightarrow{\ p|L\ } & B \end{array}$$

and thus we have a foliated bundle $\xi_L = (\hat{M},\hat{p},L)$ over L . This bundle has the following three properties:

(1) There is a distinguished section $s_L : L \longrightarrow \hat{M}$ and $\hat{L} = s_L(L)$ is a leaf of the transverse foliation $\hat{\mathcal{F}}$ on $\hat{M}$.

(2) $\hat{\mathcal{F}}$ is given by a foliated cocycle with values in $\mathrm{Diff}^r(F,y_o)$, the group of C^r diffeomorphisms of the fibre F keeping the base point y_o ($\in \hat{L} \cap F$) fixed.

(3) The holonomy representation

$$H : \pi_1(L,y_o) \longrightarrow \mathrm{Diff}^r(F,y_o)$$

of ξ_L coincides with the holonomy representation of L, as defined in II; 1.2.10, 1.2.11.

ii) Now taking an arbitrary open neighbourhood E of the section leaf $\hat{L}$, for instance a tubular neighbourhood when ξ_L is differentiable, we get a tuple

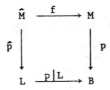

$$\psi_L = (E, p_L, \hat{L}, F_L)$$

where F_L is the foliation on E which is induced by $\hat{F}$ and p_L is the restriction of $\hat{p}$ to E .

We refer to ψ_L as a regularly foliated pseudobundle associated to the leaf L of F .

Keep in mind that $\hat{L} \in F_L$ and that F_L is transverse to the fibres of p_L. Usually we identify $\hat{L}$ and L by p_L .

Moreover, if C is the foliated cocycle in (2) defining the foliation $\hat{F}$, then F_L is given by the restriction of C to E which is a foliated cocycle with values in $H^r(F,y_o)$, the pseudogroup of local C^r diffeomorphisms of F which are defined in an open neighbourhood of y_o and which keep y_o fixed.

Similarly, the holonomy representation

$$H_{\psi_L} : \pi_1(L,y_o) \longrightarrow H^r(F,y_o)$$

of ψ_L is defined by restricting the elements of $\mathrm{im}\, H \subset \mathrm{Diff}^r(F,y_o)$ to suitable open neighbourhoods of y_o ($= p_L^{-1}(y_o) \cap L$) in F .

iii) However, there is no canonical choice for the total space E of ψ_L. Also for certain constructions we must allow E to be shrunk to a smaller open neighbourhood of L . In other words, we are interested in the germ of ψ_L near L which is called the regularly foliated microbundle associated to the leaf L of F . It is denoted by μ_L. (The general definition of a regularly foliated microbundle will be given in 1.2.7).

Let

$$\pi : H^r(F,y_o) \longrightarrow G^r(F,y_o)$$

be the natural projection onto the corresponding group of germs at y_o . Then the holonomy representation

$$H_L : \pi_1 L \longrightarrow G^r(F, y_o)$$

of μ_L is defined to be $H_L = \pi \circ H_{\psi_L}$ $(= \pi \circ H)$; cf. 1.3.3, 1.3.4, 2.1.6. Note that H_L depends only on $\mathcal{F}$ and F, but not on the choice of ψ_L.

1.2. *Generalities on foliated microbundles.*

We want to see to what extent the above considerations still hold for arbitrary foliations. First let us make precise the notions of a regularly foliated pseudobundle and a regularly foliated microbundle. For simplicity we restrict ourselves to manifolds without boundary, the alterations for bounded manifolds being obvious.

1.2.1. - *Definitions and remarks.* - i) A foliated pseudobundle of class C^r over (the ℓ-dimensional manifold) L consists of

(1) a C^r submersion $p : E \to L$,

(2) a C^r foliation $\mathcal{F}$ on E of dimension ℓ which is transverse to the fibres of p.

Furthermore, a foliated pseudobundle $\psi = (E, p, L, \mathcal{F})$ is called regularly foliated if there exists a section s of p (i.e. $p \circ s = id_L$) such that $s(L)$ is a leaf of $\mathcal{F}$. By means of s we consider henceforth L as a leaf of $\mathcal{F}$.

ii) Say that a foliated pseudobundle $\psi = (E, p, L, \mathcal{F})$ is of rank n if the fibres of p are all diffeomorphic to $\mathbb{R}^n$.

iii) For example,

$$pr_1 : L \times \mathbb{R}^n \to L ,$$

together with the horizontal foliation on $L \times \mathbb{R}^n$ and $s : L \to L \times \{0\}$ as section of pr_1, is a regularly foliated pseudobundle. It is called the product pseudobundle of rank n over L.

iv) A fibre bundle $\xi = (E, p, L)$ with ℓ-dimensional transverse foliation $\mathcal{F}$ on E need not be a foliated bundle; see II; 2.2.7 and

2.2.9. But $\psi = (E,p,L,F)$ is always a foliated pseudobundle.

v) In contrast to regularly foliated pseudobundles, we could also study foliated pseudobundles admitting a section whose image is not necessarily a leaf of the transverse foliation F. Indeed, this more general type of foliated pseudobundle is of great importance in the quantitative theory of foliations. However, it will not play any rôle in this book. Thus all foliated pseudobundles appearing henceforth are supposed to be regularly foliated. From now on we omit the word "regularly" (and sometimes also the word "foliated"). This will cause no confusion. See however 1.3.10.

1.2.2. - Definitions. i) Let $\psi = (E,p,L,F)$ and $\psi' = (E',p',L',F')$ be foliated pseudobundles of class C^r. A C^r map $f : E \to E'$ is a map of foliated pseudobundles if

(1) f preserves the fibres, i.e. f induces $\bar{f} : L \to L'$ such that $p' \circ f = \bar{f} \circ p$,

(2) for every $b \in L$, the restriction of f to the fibre $p^{-1}(b)$ is a diffeomorphism of $p^{-1}(b)$ onto an open subset of $(p')^{-1}(\bar{f}(b))$,

(3) f preserves the foliations, i.e. f is transverse to F' and $F = f^*F'$, and $f(L) = L'$.

Isomorphisms between foliated pseudobundles over the same base are defined in the obvious way.

ii) A (foliated) sub-pseudobundle of $\psi = (E,p,L,F)$ is of the form (E_o,p_o,L,F_o) where E_o is an open subset of E containing L, $p_o = p|E_o$ and $F_o = F|E_o$.

Sub-pseudobundles of the product pseudobundle are called trivial.

1.2.3. - Lemma. - Let $\psi = (E,p,L,F)$ be a trivial pseudobundle over a compact manifold L. Then ψ contains a product pseudobundle.

Proof : Take a finite covering of L by bidistinguished open cubes $U_1,\ldots,U_s$ such that $U_{i-1} \cap U_i \neq \emptyset$ and $U_i \cap L \neq \emptyset$. Then proceed by induction on s. □

Simple examples show that the compactness of L is essential in this lemma.

Any foliated pseudobundle contains a sub-pseudobundle which is of rank n . For the proof of this assertion in 1.2.5 we introduce a special sort of open covering for foliated manifolds.

1.2.4. - Definition. - Let (M,F) be a foliation. We say a denumerable covering $\mathcal{U} = \{U_i\}_{i\in\mathbb{N}}$ of (M,F) by distinguished sets is nice if the following conditions hold :

(1) The covering $\mathcal{U}$ is locally finite.

(2) The U_i are distinguished open cubes.

(3) When $U_i \cap U_j \neq \emptyset$ there is a distinguished open cube U_{ij} (not necessarily belonging to $\mathcal{U}$) such that $\mathrm{cl}(U_i \cup U_j) \subset U_{ij}$.

Similarly, if there is a transverse foliation $F^{\pitchfork}$ to F (of complementary dimension) then we speak of a nice covering $\{U_i\}_{i\in\mathbb{N}}$ of $(M,F,F^{\pitchfork})$ (by bidistinguished open cubes). This means that $\{U_i\}$ consists of bidistinguished open cubes subject to conditions (1) - (3).

It is not hard to see that nice coverings of (M,F) (resp. $(M,F,F^{\pitchfork})$) always exist. Moreover, one can show that every foliated cocycle has a nice refinement. Consequently, we may suppose that any foliation is given by a foliated cocycle whose underlying covering is nice.

1.2.5. - Lemma. - Let $\psi = (E,p,L,F)$ be a foliated pseudobundle with fibre dimension n . Then ψ contains a sub-pseudobundle which is of rank n .

Proof : Let $\{U_i\}_{i\in\mathbb{N}}$ be a nice covering of E by bidistinguished open cubes, with distinguished maps $f_i : U_i \to I^n = [-1,1]^n$ with respect to F. Possibly after shrinking E in the fibre direction, we may assume that each U_i intersects L in exactly one plaque P_i and that $f_i(P_i) = 0$.

Starting from $\{U_i\}$, we construct recursively a new covering $\{U_i'\}$ of L by bidistinguished cubes as follows.

(1) Put $U_1' = U_1$.

(2) If U_i' is already defined for $i \leqslant s-1$, then we choose an open cube $Q_s \subset \bigcap_{i=1}^{s} f_s(U_i')$ centered at 0 and put $U_s' = f_s^{-1}(Q_s)$.

Now if $E_s = \bigcup_{i=1}^{s} U_i'$ and $p_s = p|E_s$ then $p_{s-1}^{-1}(x) = p_s^{-1}(x)$ for $x \in L \cap E_{s-1} \cap E_s$. We conclude that $E_o = \bigcup_s E_s$ has the required properties. □

Our interest in nice coverings is also based on the following property which will be used in section 2.2.

1.2.6. - Lemma. - <u>Let</u> $C = (\{(U_i,f_i)\},\{g_{ij}\})$ <u>be a foliated cocycle whose underlying covering is nice. Then each</u> g_{ij} <u>determines a local diffeomorphism</u> $\tilde{g}_{ij} : f_j(U_j) \to f_i(U_i)$ <u>such that</u> $\tilde{g}_{ij}(f_j(x)) = g_{ij}(x)(f_j(x))$ <u>for any</u> $x \in U_i \cap U_j$.

Proof : Let U_{ij} be a distinguished open cube of the foliation given by C such that $cl(U_i \cup U_j) \subset U_{ij}$. Since $U_i \cap U_{ij} = U_i$ is connected, the coordinate transformation g_i with respect to U_{ij} and U_i is constant, and similarly for U_{ij} and U_j. Thus $\tilde{g}_{ij} = g_i^{-1} \circ g_j$ has the required property. □

1.2.7. - Definition.- Let $\psi = (E,p,L,F)$ be a foliated pseudo-bundle (of class C^r and fibre dimension n). The family of all sub-pseudobundles of ψ (i.e. the <u>germ of</u> ψ <u>near</u> L) is referred to as a <u>foliated microbundle (of class</u> C^r <u>and rank</u> n) <u>over</u> L, denoted μ.

The microbundle corresponding to the product pseudobundle $L \times \mathbb{R}^n$ is called the *trivial microbundle* (of rank n) over L.

Maps and *isomorphisms* of foliated microbundles are defined via representatives.

1.2.8. - Induced foliated pseudobundles and microbundles are constructed by analogy to II; 1.1.12. Let $\psi = (E,p,L,F)$ be a foliated pseudobundle and $\bar{f} : L' \to L$ a C^r map. We set

$$E' = \{(b',x) \in L' \times E \mid \bar{f}(b') = p(x)\}$$

and get a commutative diagram

$$\begin{array}{ccc} E' & \xrightarrow{f} & E \\ {\scriptstyle p'}\downarrow & & \downarrow{\scriptstyle p} \\ L' & \xrightarrow{\bar{f}} & L \end{array} \quad ,$$

where p' and f are the canonical projections. It is easily seen that f is transverse to F, thus E' is equipped with a foliation $F' = f^*F$ which makes $\psi' = \bar{f}^*\psi = (E',p',L',F')$ a foliated pseudobundle over L' and f a pseudobundle map.

The germ of ψ' near L' is well-defined and is called the *(foliated) microbundle induced by the map* $\bar{f}$. It is denoted by $\bar{f}^*\mu$.

To begin with we state two results on microbundles over the disk and over simply connected base spaces. These results will be used later.

1.2.9. - Lemma.- *Every foliated microbundle* μ *over* $I \times I$ *is trivial. More precisely, any representative of* μ *contains a product pseudobundle.*

Proof : Starting from an arbitrary representative $\psi = (E,p,I \times I,F)$ of μ , we want to find $E_o \subset E$ containing $I \times I$

such that $(E_o, p|E_o, I \times I, F|E_o)$ is trivial.

Let $\{U_i\}$, $i = 1,\dots,k$, be a finite covering of $L \subset E$ by bidistinguished open cubes such that the plaques $P_i = U_i \cap L$ cover $I \times I$ as indicated in figure 1. Denote by V_i the intersection of $p^{-1}(0)$ with the saturation of U_i, (i.e. with the smallest F-saturated subset of E containing U_i). Then it follows by induction on i (proceeding as indicated by the arrows in fig. 1) that the saturation E_o of $\bigcap_{i=1}^{k} V_i$ has the required properties. □

Figure 1

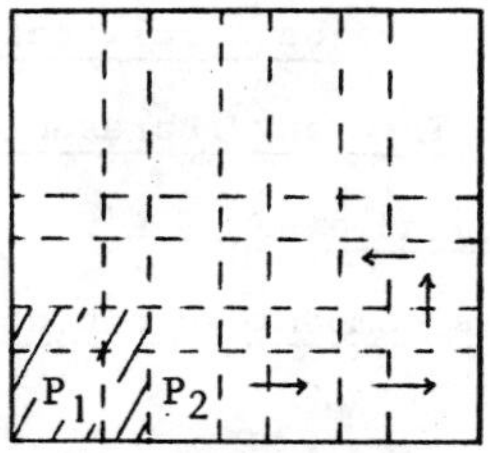

$I \times I$

Notice that a foliated pseudobundle over the disk need not be trivial. Here is an example. Let E be the space $D^2 \times [-1,1]$ with $\{0\} \times [\frac{1}{2},1]$ removed, and p the projection onto D^2. To construct the transverse foliation F we start from the suspension $(\partial D^2 \times [-1,1], F_f)$ of a diffeomorphism $f \in \mathrm{Diff}^r_+([-1,1])$ such that

$$f(x) \begin{cases} = x & \text{for } x \leqslant \frac{1}{2} \\ < x & \text{for } \frac{1}{2} < x < 1 . \end{cases}$$

Then F_f extends in an obvious way to a foliation F on E which is transverse to the fibres of p ; cf. fig. 2. If D^2 is identified with $D^2 \times \{0\}$ then (E,p,D^2,F) is non-trivial.

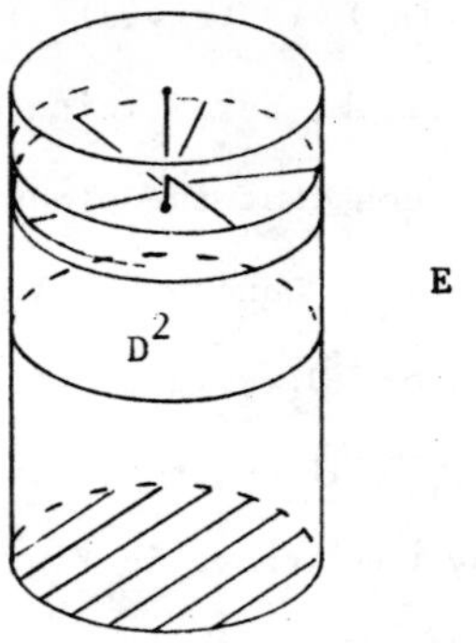

Figure 2

1.2.10. - *Lemma*.- Let μ be a foliated microbundle over a simply connected manifold L . Then any representative (E,p,L,F) of μ contains a trivial sub-pseudobundle $E_o \subset E$.

Moreover, when L is compact we can choose E_o to be a product.

Proof : Let $J_s = \{1,\dots,s\}$ be a set of indices. By a simple cycle in J_s we mean a map $j : \{0,\dots,q\} \to J_s$ such that $j_k = j_\ell, k \neq \ell$, if and only if $k = 0$ and $\ell = q$. Evidently we then have $q \leqslant s$, hence there is only a finite number of simple cycles in J_s .

a) Now suppose that L is compact and is covered by a family $\{U_j\}_{j\varepsilon J_s}$ of bidistinguished open cubes with $E = \bigcup_{j\varepsilon J_s} U_j$. We assume that there exists a sequence of open neighbourhoods

$$E_1 \supset E_2 \supset \dots \supset E_k \supset \dots$$

of L in E such that $L = \bigcap_k E_k$ and no E_k is trivial.

Since there is only a finite number of simple cycles in J_s , there exists a cycle $(j_o,\dots,j_q)$ and for each k a sequence of F-plaques $\{P_o^k,\dots,P_q^k\}$ such that

(1) P_ℓ^k is a plaque of U_{j_ℓ} , but $P_o^k \neq P_q^k$,

(2) $P_{\ell-1}^k \cap P_\ell^k \neq \emptyset$ for each $\ell \varepsilon \{1,\dots,q\}$.

Now let c be a loop in L which is the image under p of a path in E_1 joining P_o^1 to P_q^1 and contained in $\bigcup_{\ell=o}^{q} P_\ell^1$. Since c is homotopic to zero, we get a contradiction to 1.2.9. Hence one of the E_k must be trivial. It was shown in 1.2.3 that E_k contains a product E_o.

b) When L is non-compact, it is the union of an exhausting sequence of compact submanifolds

$$L_1 \subset L_2 \subset \ldots L_j \subset \ldots \quad .$$

Let E_o^j be a product sub-pseudobundle of $p|p^{-1}(L_j)$, according to a). Then $E_o = \bigcup_j E_o^j$ is the required trivial neighbourhood of L in E. □

1.2.11. - Definition und remark. - The germ $g(F,L)$ of an arbitrary foliation F near an arbitrary leaf $L \in F$, as well as the notion of homeomorphism (diffeomorphism, etc.) between germs, is defined just as for circle leaves in I; 3.2.2. Diffeomorphisms between germs $g(F,L)$ and $g(F',L)$ near the same leaf L are called isomorphisms.

For foliated pseudobundles over the same leaf L we have two notions of isomorphism between germs near L. These are related in the following way showing that to establish an isomorphism between micro-bundles we need not worry about the submersions. More precisely, we have:

1.2.12. - Lemma. - Let $\psi = (E,p,L,F)$ and $\psi' = (E',p',L,F')$ be foliated pseudobundles over L. If the germs of F and F' near L are isomorphic by a diffeomorphism which is the identity on L, then the foliated microbundles represented by ψ and ψ' are isomorphic.

Proof : Assume we are given a diffeomorphism

$$h : U \to U'$$

between open neighbourhoods of L in E and E', respectively, such that $h^*(F'|U') = F|U$ and $h|L = id$. We want to find a local diffeomorphism

$$\hat{h} : E \to E'$$

defined in some open neighbourhood of L in U which takes $p^{-1}(x)$ to $(p')^{-1}(x)$ for each $x \in L$, and such that $\hat{h}^*(F') = F$.

For each $x \in L$ we can find bidistinguished open cubes Q and Q' around x in U and U', respectively, such that $h(Q) \subset Q'$. For $y \in Q$ we define $\hat{h}(y)$ to be the intersection of the plaque of Q' through the point $h(y)$ and the fibre of p' over $p(y)$; see fig. 3. Using that $h|L = id$, it is not hard to see that this gives a well-defined diffeomorphism on some open neighbourhood of L in U which preserves the foliations F and F' and takes $p^{-1}(x)$ to $(p')^{-1}(x)$, for each $x \in L$. □

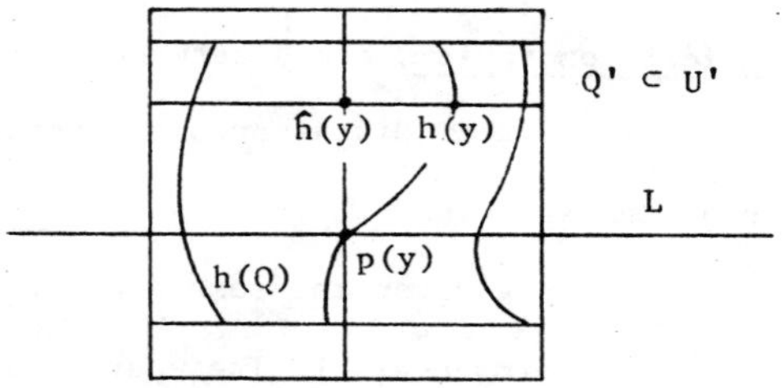

Figure 3

1.3. *Holonomy of foliated microbundles.*

Our next purpose is to define the holonomy representation of a foliated microbundle. We proceed by analogy to II; 1.3.5 - 7.

The holonomy representation H_μ of a foliated microbundle μ of class C^r and rank n will have values in $G_n^r(0)$, the group of germs at 0 of local C^r diffeomorphisms of $\mathbb{R}^n$ which are defined in a neighbourhood of 0 and keep 0 fixed. As a first step we want to find "representatives" of H_μ, with values in the pseudogroup $H_n^r(0)$ of representatives of elements of $G_n^r(0)$. For that we first introduce a special class of representatives of μ.

1.3.1. - *Definition*.- Let $\psi = (E,p,L,F)$ be a foliated pseudobundle and let $q : \tilde{L} \to L$ be the universal covering of L. We say that ψ is <u>adapted</u> if the induced pseudobundle $q^*\psi$ over $\tilde{L}$ is trivial.

1.3.2. - *Lemma*.- i) <u>Every foliated pseudobundle</u> $\psi = (E,p,L,F)$ <u>contains a sub-pseudobundle which is adapted</u>.

ii) <u>If</u> $\psi = (E,p,L,F)$ <u>is adapted and</u> $u : K \to L$ <u>is continuous, with</u> K <u>simply connected, then</u> $u^*\psi$ <u>is trivial</u>.

Proof : By 1.2.10, the induced pseudobundle $q^*\psi$ over $\tilde{L}$ contains a trivial sub-pseudobundle $(\tilde{E},\tilde{p},\tilde{L},\tilde{F})$. We set $E_o = f(\tilde{E})$, where f is the induced pseudobundle map. Then $\psi_o = (E_o, p|E_o, L, F|E_o)$ is adapted.

The proof of ii) is trivial. □

1.3.3. - *Remarks and definitions*.- (Construction of the holonomy representation). Let μ be a foliated microbundle of class C^r and rank n which is represented by $\psi = (E,p,L,F)$.

i) The restriction of p to a leaf $L' \neq L$ is locally diffeomorphic but (in contrast to foliated bundles) in general not a covering map. It might not even be surjective.

ii) a) Let $c : [0,1] \to L$ be a path from b_o to b_1. We denote by F_o and F_1 the fibres of p over b_o and b_1, respectively. Then the set of points $y \in F_o$ such that there is a lifting $\tilde{c}_y$ of c from y to $\tilde{c}_y(1) \in F_1$ in the leaf through y is non-empty and open. Since all these lifts are unique, we get a well defined local C^r diffeomorphism

$$T_c : F_o \to F_1 ,$$

the <u>local translation of</u> F_o <u>to</u> F_1 <u>along</u> c.

b) Let $\bar{f} : I \times I \to L$ be a homotopy between two paths c and

c' with endpoints b_o and b_1. By 1.2.9, the induced microbundle $\bar{f}^*\mu$ is trivial. This implies immediately that the germ of T_c at 0 depends only on the homotopy class of c .

However, as was observed when pseudobundles over $I \times I$ were studied (compare 1.2.9 and the example following it), the local translation T_c depends on the path c . In other words, there are nullhomotopic paths c with $T_c \neq id$ (but with the germ at 0 being always the identity).

In order to remedy this phenomenon, we have to restrict ourselves to adapted representatives of μ .

c) Suppose that ψ is adapted and has rank n . Then if c is nullhomotopic, it follows that $T_c = id$. Thus if c and c' are homotopic paths in L with endpoints b_o and b_1 then T_c and $T_{c'}$ coincide on the intersection of their domains. Therefore to the homotopy class γ of c there is associated a well-defined local diffeomorphism

$$T_\gamma : F_o \to F_1 ,$$

the <u>local translation along</u> γ , whose domain is the union of the domains of the local maps T_c , where c is a representative of γ .

d) Now taking $b_1 = b_o$ and fixing a parameterization of $(F_o, F_o \cap L)$ as open neighbourhood of $0 \in \mathbb{R}^n$, we have assigned to each $\gamma \in \pi_1(L,b_o)$ an element $T_\gamma \in H_n^r(0)$. Clearly $T_{\gamma\gamma'} = T_{\gamma'} \circ T_\gamma$ where both sides are defined. Therefore, if we define

$$H_\psi : \pi_1(L,b_o) \to H_n^r(0)$$

by $H_\psi(\gamma) = T_{\gamma^{-1}}$ then we get a <u>homomorphism of pseudogroups</u>, (i.e. $H_\psi(1) = id$, $H_\psi(\gamma^{-1}) = H_\psi(\gamma)^{-1}$ and $H_\psi(\gamma\gamma') = H_\psi(\gamma) \circ H_\psi(\gamma')$ on the intersection of domains).

This homomorphism is called the <u>holonomy (representation) of the foliated pseudobundle</u> ψ . It is defined up to conjugation with an element of $H_n^r(0)$ depending on the parameterization of F_o .

Let

$$\pi : H^r_n(0) \to G^r_n(0)$$

be the natural projection.

1.3.4. - Definition.- (Notation as in 1.3.3) The group homomorphism

$$H_\mu = \pi \circ H_\psi : \pi_1(L,b_o) \to G^r_n(0)$$

is called the holonomy (representation) of the foliated microbundle μ . It is defined up to conjugation with an element of $G^r_n(0)$. This definition does not depend on the choice of the adapted representative ψ of μ .

The following results illustrate the close relationship between foliated microbundles and foliated bundles.

1.3.5. - Proposition.- Let $\psi = (E,p,L,F)$ be adapted. The following two conditions are equivalent.

i) ψ is trivial.

ii) The holonomy of ψ is trivial.

Proof : Evidently i) implies ii).

If H_ψ is trivial then it follows from 1.3.3 that a leaf of F intersects a fibre of p in at most one point. We conclude that ψ must be trivial. □

1.3.6. - Corollary.- (Cf. II; 1.3.8) A foliated microbundle is trivial if and only if its holonomy representation is trivial.

1.3.7. - Theorem.- (Cf. II; 1.2.9 and II; 1.3.1) Two foliated microbundles μ and μ' of class C^r and rank n over L are C^r isomorphic if and only if their holonomy representations are conjugate in $G^r_n(0)$.

Proof : We have to prove the "if" statement, the "only if" part being evidently true. After possibly changing the parameterization, we may suppose that the holonomy representations H_μ and $H_{\mu'}$ are equal. We choose adapted representatives of μ and μ' and pass to the induced pseudobundles over the universal covering $\tilde{L}$ of L . This gives us two trivial pseudobundles which may be canonically identified with sub-pseudobundles of $\tilde{L} \times \mathbb{R}^n$. On the intersection of these two sub-pseudobundles of $\tilde{L} \times \mathbb{R}^n$ the identity induces an isomorphism between representatives of μ and μ' . □

In order to complete the analogy between foliated microbundles and foliated bundles we want to describe briefly the suspension construction for foliated microbundles. This construction occured already implicitly in the proof of the last theorem. It consists of assigning to a homomorphism $H : \pi_1 L \to G_n^r(0)$ a foliated microbundle μ over L whose corresponding holonomy equals H . We then say that μ is the suspension of H .

1.3.8. - Theorem.- Let L be a manifold and $\hat{H} : \pi_1 L \to H_n^r(0)$ a homomorphism of pseudogroups. Then there exists a foliated microbundle μ of class C^r and rank n over L such that the holonomy representation of μ equals $\pi \circ \hat{H} : \pi_1 L \to G_n^r(0)$.

Proof : Let $q : \tilde{L} \to L$ be the universal covering of L . By means of $\hat{H}$ we define a "local action" of $\pi_1 L$ on $\tilde{E} = \tilde{L} \times \mathbb{R}^n$ as follows. For each $\gamma \in \pi_1 L$ we have a local C^r diffeomorphism

$$g : \tilde{E} \longrightarrow \tilde{E}$$
$$(\tilde{x}, y) \mapsto (\gamma(\tilde{x}), \hat{H}(\gamma)(y))$$

which is defined on a neighbourhood of $\tilde{L} = \tilde{L} \times \{0\}$. Here γ acts on the first coordinate as a covering translation of q .

On $\tilde{E}$ consider the relation ρ given by

$$a \rho b \text{ if and only if } b = \hat{H}(\gamma)(a) \text{ for some } \gamma \in \pi_1 L .$$

Then ρ generates an equivalence relation, again denoted ρ , by writing $a \rho b$ if there are $a_o,\dots,a_s \in \tilde{E}$, $a_o = a$, $a_s = b$, such that $a_i \rho a_{i-1}$ in the above sense, $i = 1,\dots,s$.

Let $\tilde{\pi} : \tilde{E} \to E' = \tilde{E}/\rho$ be the quotient map. If $pr : \tilde{E} \to \tilde{L}$ is the projection onto the first factor then there is an induced map

$$p : E' \to L$$

such that $p \circ \tilde{\pi} = q \circ pr$. Notice that p has a natural section, because $\tilde{L}$ is saturated under ρ and $\tilde{L}/\rho = L$.

In general, the quotient **space** E' is non-Hausdorff. This is the crucial point in the proof. But, as we shall see, the zero-section $\tilde{L}$ admits a neighbourhood W in $\tilde{E}$ which is mapped by $\tilde{\pi}$ onto a neighbourhood E of L in E' which is Hausdorff.

For the construction of W we first take an open covering $\{V_\sigma\}_{\sigma\in K}$ by small regular neighbourhoods of the simplices σ of some triangulation K of L . Furthermore, the following condition should hold:

$$V_{\sigma_1} \cap V_{\sigma_2} \begin{cases} = \emptyset & \text{if } \sigma_1 \cap \sigma_2 = \emptyset \\ \subset V_\sigma & \text{if } \sigma_1 \cap \sigma_2 = \sigma \end{cases}$$

The lifting of $\{V_\sigma\}$ under q yields an open covering $\{V_{\tilde{\sigma}}\}_{\tilde{\sigma}\in\tilde{K}}$ of $\tilde{L}$ whose elements are regular neighbourhoods of the simplices $\tilde{\sigma}$ of $\tilde{K}$, the lift to $\tilde{L}$ of the triangulation K . Clearly

$$V_{\gamma(\tilde{\sigma})} = \gamma(V_{\tilde{\sigma}}) \quad \text{for each } \tilde{\sigma} \in \tilde{K} \text{ and each } \gamma \in \pi_1 L .$$

Now the neighbourhood W of $\tilde{L} \subset \tilde{E}$ is obtained inductively as follows. We first want to find for each $\tilde{\sigma} \in \tilde{K} = \cup \tilde{K}^{(k)}$ a suitable open disk neighbourhood $D_{\tilde{\sigma}}$ of $0 \in \mathbb{R}^n$.

Beginning with $k = 0$, we pick for each vertex $\sigma \in K^{(o)}$ a vertex $\tilde{\sigma}_o \in \tilde{K}^{(o)}$ in $q^{-1}(\sigma)$ and put $D_{\tilde{\sigma}_o} = \mathbb{R}^n$. Then for each $\tilde{\sigma} \in \tilde{K}^{(o)} \cap q^{-1}(\sigma)$ there is a unique $\gamma \in \pi_1 L$ such that $\tilde{\sigma} = \gamma(\tilde{\sigma}_o)$. We choose $D_{\tilde{\sigma}}$ to be an open disk around zero in $\operatorname{im}\hat{H}(\gamma)$. Then for each $\sigma \in K^{(k)}$, $k \geq 1$, we choose $\tilde{\sigma}_k \in \tilde{K}^{(k)} \cap q^{-1}(\sigma)$ and an open disk $D_{\tilde{\sigma}_k}$ around $0 \in \mathbb{R}^n$ such that

$$D_{\tilde{\sigma}_k} \subset \bigcap_i D_{\gamma_i(\partial_i \tilde{\sigma}_k)} ,$$

where ∂_i denotes the i-th face, and $\gamma_i \in \pi_1 L$ is the unique element such that $\gamma_i(\partial_i \tilde{\sigma}_k)$ is some $\tilde{\sigma}_{k-1}$. For arbitrary $\tilde{\sigma} \in \tilde{K}^{(k)}$ the corresponding open disk $D_{\tilde{\sigma}}$ is then found in a similar way to the case $k-1$.

Now if we set $W_{\tilde{\sigma}} = D_{\tilde{\sigma}} \times V_{\tilde{\sigma}}$, for $\tilde{\sigma} \in \tilde{K}$, then

$$\tilde{\pi}(W_{\tilde{\sigma}_k}) = \tilde{\pi}\Big(\bigcup_{\gamma \in \pi_1 L} W_{\gamma(\tilde{\sigma}_k)} \Big) ,$$

and $\tilde{\pi}(W_{\tilde{\sigma}_k})$ and $W_{\tilde{\sigma}_k}$ are homeomorphic. On the other hand, it is easily seen that points in E' which cannot be separated by open sets must lie in the same fibre of p . Thus if $W = \bigcup_{\tilde{\sigma} \in \tilde{K}} W_{\tilde{\sigma}}$ then $E = \tilde{\pi}(W)$ is Hausdorff.

Denote by F the foliation on E which is induced by the horizontal foliation on $\tilde{L} \times \mathbb{R}^n$. Then the foliated microbundle μ over L represented by (E,p,L,F) has the required properties. □

1.3.9. - <u>*Remark*</u>.- Of course, it would be better in the last theorem to start from a homomorphism $H : \pi_1 L \to G_n^r(0)$ and then construct a lift $\hat{H} : \pi_1 L \to H_n^r(0)$ of H . We do not know, however, whether any homomorphism H can be lifted to a homomorphism of pseudogroups $\hat{H}$.

The preceding construction will not be used later.

1.3.10.- Definitions and exercises.- (Haefliger structures)

i) A C^r Haefliger cocycle (H - cocycle, for short) of codimension n on the manifold M is a pair $C = (\{(U_i, f_i)\}, \{g_{ij}\})$ where $\{U_i\}$ is an open covering of M,

$$f_i : U_i \to \mathbb{R}^n$$

is a (not necessarily submersive) C^r map and for $U_i \cap U_j \neq \emptyset$ the maps

$$g_{ij} : U_i \cap U_j \to H_n^r$$

are locally constant and satisfy:

(1) $f_i(x) = g_{ij}(x)(f_j(x))$, for $x \in U_i \cap U_j$.

(2) For $x \in U_i \cap U_j \cap U_k$ we have

$$g_{ik}(x) = g_{ij}(x) \circ g_{jk}(x)$$

in a neighbourhood of $f_k(x)$.

Notice that, in contrast to foliated cocycles, condition (2) is not a consequence of (1).

ii) Let $C = (\{(U_i, f_i)\}, \{g_{ij}\})$ be a C^r H - cocycle of codimension n with $\{U_i\}$ locally finite.

a) Along the lines of II; 1.1.11 construct a submersion $p : E \to M$. Hint: The difficulty is to find E to be Hausdorff.

b) Show that the horizontal foliations on $U_i \times \mathbb{R}^n$ induce a foliation F of codimension n on E which is transverse to the fibres of p and whose transverse structure is given by $\{g_{ij}\}$.

c) The local sections $U_i \to U_i \times \mathbb{R}^n$, $x \mapsto (x, f_i(x))$ induce a section $s : M \to E$ of p .

We say that $\psi_C = (E, p, s(M), F)$ is a Haefliger pseudobundle (of class C^r) over M (defined by the cocycle C).

The germ of ψ_C near $M = s(M)$ is called a Haefliger micro-bundle over M (of rank n when n is the fibre dimension).

iii) Introduce the notion of equivalence between H - cocycles similarly to that for ordinary cocycles.

An equivalence class of C^r H - cocycles of codimension n is called a C^r Haefliger structure (H - structure) of codimension n on M.

Note that every C^r foliation on M determines a C^r H - structure on M in a canonical way. On the other hand, every manifold M admits a trivial H - structure of any codimension.

a) Every H - cocycle is equivalent to one whose underlying covering is locally finite.

b) Using the obvious notion of isomorphism of Haefliger micro - bundles, show that there is a 1 - 1 correspondence between the isomorphism classes of C^r H - microbundles of rank n over M and the C^r H - structures of codimension n on M .

iv) Let $\psi = (E,p,L,F)$ be a Haefliger pseudobundle, defined by the H - cocycle $C = (\{(U_i,f_i)\},\{g_{ij}\})$. In general $M = s(M)$ is not a leaf of F .

a) M is a leaf of F if and only if all f_i are constant (U_i being connected).

b) C is a foliated cocycle if and only if M is transverse to F .

c) If $s : M \hookrightarrow E$ is transverse to F then s^*F is the foliation given by C .

2. *Holonomy of leaves.*

In this paragraph we apply the results of the preceding sections to define the holonomy of leaves of arbitrary foliations. Our approach here seems to be more "geometric and global" than that usually found in the literature.

We first associate to each leaf L of a foliation $(M,\mathcal{F})$ a foliated pseudobundle $\psi_L = (E,p,L,\mathcal{F}_L)$ over L together with a homomorphism of foliated manifolds $\alpha : E \to M$ which on L is the natural inclusion. The holonomy representation of L is then defined to be the holonomy of the foliated microbundle represented by ψ_L.

An essential advantage of the approach taken here is that the theorems of Haefliger (see 2.1.7) and Reeb (see 2.1.8) can be derived fairly easily.

Also we give in 2.2 a description of holonomy using foliated cocycles (thus following the usual way to define holonomy). Moreover, it is shown that in any foliated manifold almost all leaves have trivial holonomy (see 2.2.6).

2.1. *Unwrapping of leaves ; leaf holonomy.*

Now let L be a leaf of the foliated manifold $(M,\mathcal{F})$ of class C^r and codimension n. We associate to L a foliated pseudobundle in the following way.

2.1.1. - *Proposition.*- There exists a foliated pseudobundle $\psi_L = (E,p,L,\mathcal{F}_L)$ of class C^r and rank n over L and a map

$$\alpha : E \to M$$

such that

(1) α is a C^r immersion, i.e. each point of E has a neighbourhood which is mapped by α diffeomorphically onto its image,

(2) $\alpha|L$ is the inclusion,

(3) α is transverse to $\mathcal{F}$ and $\alpha^*\mathcal{F} = \mathcal{F}_L$.

Furthermore, the microbundle μ_L represented by ψ_L is unique in the following sense. If (ψ_L',α') is a similar pair satisfying (1) - (3) then μ_L and μ_L' are isomorphic foliated microbundles.

Proof : First suppose $r \geq 1$. Then, with respect to a riemannian metric on M , we may identify the normal bundle $\nu_L = (N,q,L)$ of L with the orthogonal complement in TM of the restriction to L of the tangent bundle of F .

If α denotes the exponential map then there exists an open neighbourhood E of the zero - section L in N such that (1) and (2) hold. Moreover, E can be chosen so that the fibres of $p = q|E$ are disks whose images under α are transverse to F .

Thus, if we put $F_L = \alpha^*(F|\mathrm{im}\,\alpha)$ then $\psi_L = (E,p,L,F_L)$ and α fulfill conditions (1), (2), (3).

It was proved by Siebenmann in [Si] and Harrison in [Har] that also in the case $r = 0$ there exists ψ_L and α as required. The proof uses the topological isotopy extension theorem. Once this is assumed to be given, the proof is technical but not very hard. We will not give any details on this point; see [Har; p.104], for instance.

Now assume that $\psi'_L = (E',p',L,F'_L)$ and $\alpha' : E' \to M$ satisfy (1) - (3). After possibly shrinking E and E' in the fibre direction, we can find nice coverings $\{U_i\}_{i\in\mathbb{N}}$ and $\{U'_i\}_{i\in\mathbb{N}}$ of E and E' , resp., by bidistinguished open cubes with the following additional properties:

(1) If $U_i \cap U_j \neq \emptyset$ then there exists a bidistinguished open cube U_{ij} with $\mathrm{cl}(U_i \cup U_j) \subset U_{ij}$ such that the restriction of α to U_{ij} is injective (and similarly for $\{U'_i\}$).

(2) $P_i = U_i \cap L \neq \emptyset$ (resp. $P'_i = U'_i \cap L \neq \emptyset$) and $P_i \subset P'_i$ for each i .

We define a map $h : E \to E'$ by setting $h(x) = (\alpha'_i)^{-1} \circ \alpha_i(x)$, where $x \in U_i$ and α_i and α'_i are the restrictions of α and α' to U_i and U'_i , respectively. Then h is a well-defined foliation preserving C^r map. Furthermore, h induces a diffeomorphism of the germs of F_L and F'_L near L . To get an inverse to h just reverse the rôles of ψ_L and ψ'_L . An application of 1.2.12 completes the proof. □

2.1.2. - *Remark and definition*.- We observe that, in the previous proposition, the images of the fibres of p under the map α in general do not constitute a foliation on $\alpha(E)$. However, if L is a proper leaf, i.e. its manifold topology is the same as that induced by the topology of the surrounding manifold M (in other words L is embedded in M), then we can choose E so that α becomes one - to - one. More precisely, we have:

2.1.3. - *Corollary*.- Suppose that L is a proper leaf of (M,F). Then we can find $\psi_L = (E,p,L,F_L)$ and $\alpha : E \to M$ in 2.1.1 so that α is a C^r diffeomorphism onto a neighbourhood V of L in M .

In particular, the foliation F restricted to V admits a transverse foliation (of complementary dimension).

2.1.4. - *Definition and remarks*.- Let L be a leaf of the foliation F .

i) Any foliated pseudobundle $\psi_L = (E,p,L,F_L)$ over L provided by proposition 2.1.1 is referred to as an unwrapping (pseudobundle) of F near L ; the foliated microbundle μ_L represented by ψ_L is called the unwrapping microbundle of F near L .

ii) Note that the unwrapping microbundle is already determined by the restriction of F to an arbitrary open neighbourhood of L in M .

iii) In case L is proper the germs of F_L and F near L can be identified.

iv) For leaves of foliated bundles the unwrapping construction is just the localization; see 1.1.

2.1.5. - *Examples*.- i) Let (Σ,F) be a foliated surface and $L \in F$ a circle leaf. The unwrapping microbundle μ_L is represented by a foliated pseudobundle over L given by

(1) a small open neighbourhood U of L in Σ ,

(2) the projection of U onto L along the leaves of an arbitrary foliation transverse to $\mathcal{F}$,

(3) the foliation $\mathcal{F}_L = \mathcal{F}|U$.

On the other hand, let $\mathcal{F}$ ba a foliation on the torus T^2 defined by suspension of a Denjoy diffeomorphism of S^1 . We have the commutative diagram (cf. I; 3.1.2)

$$\begin{array}{ccc} S^1 \times \mathbb{R} & \xrightarrow{pr} & \mathbb{R} \\ \pi \downarrow & & \downarrow q \\ T^2 & \xrightarrow{p} & S^1 \end{array} .$$

For any leaf $L \in \mathcal{F}$, the foliated microbundle represented by the product pseudobundle $(S^1 \times \mathbb{R}, pr, \mathbb{R})$, together with $\alpha = \pi$, is an unwrapping microbundle of $\mathcal{F}$ near L . This shows that the leaves of $\mathcal{F}$ (which are all canonically diffeomorphic to $\mathbb{R}$) have isomorphic unwrapping micro-bundles, although there are two different kinds of leaves in $\mathcal{F}$, namely

a) the proper leaves for which α can be chosen to be injective,

b) the exceptional leaves for which α is never injective.

ii) Let $(D^{m-1} \times S^1, \mathcal{R})$ be a Reeb component and let $L \in \mathcal{R}$ be a plane leaf. We think of $\mathcal{R}$ as obtained by the equivariant submersion

$$D^{m-1} \times S^1 \xleftarrow{\pi} H^m_* \xrightarrow{pr_m} \mathbb{R} ;$$

cf. II; 1.4.4. If $\tilde{L}$ is a leaf in the simple foliation $\mathcal{F}_o$ on H^m_* defined by pr_m , with $\pi(\tilde{L}) = L$, then there is a saturated neighbourhood U of $\tilde{L}$ in H^m_* such that

$$\alpha = \pi|U : (U, \mathcal{F}_o|U) \to (D^{m-1} \times S^1, \mathcal{R})$$

is an $\mathcal{F}$-isomorphism onto its image. We conclude that μ_L is trivial. See also II; 1.4.3 and the next exercises.

By means of the unwrapping microbundle we are now able to apply the results of section 1.2 to the definition of holonomy for leaves of arbitrary foliations.

2.1.6. - *Definitions and remarks*.- i) Let $\mathcal{F}$ be a foliation of class C^r and codimension n and let L be a leaf of $\mathcal{F}$.

The holonomy (representation) of L,

$$\text{hol} : \pi_1 L \to G_n^r(0) \qquad ,$$

is by definition the holonomy representation of the unwrapping microbundle μ_L of $\mathcal{F}$ near L. (By 1.3.3 and 2.1.1, this group homomorphism is well-determined up to conjugation).

The image $\text{hol}(L)$ of hol is called the holonomy group of L.

ii) In codimension one we have also the notion of one-sided holonomy. More precisely, if L is a boundary leaf of $(M,\mathcal{F})$ or a two-sided leaf in the interior of M, we may define the holonomy (resp. right or left holonomy) of L to be a representation of $\pi_1 L$ in the group $G^r(\mathbb{R}^+,0)$. This is all done in a straightforward way.

iii) (See II; 1.2.10 and 1.2.11) For a leaf L of a foliated bundle (with base point $x \in L$) the holonomy H_x of L was defined to be a homomorphism of $\pi_1(L,x)$ in the group $\text{Diff}^r(F,y_o)$ of C^r diffeomorphisms of the fibre F keeping some point $y_o \in F$ fixed. When

$$\pi : \text{Diff}^r(F,y_o) \to G^r(F,y_o)$$

is the natural projection onto the corresponding group of germs, we get a group homomorphism

$$\pi \circ H_x : \pi_1 L \to G^r(F,y_o) .$$

Finally, identifying some neighbourhood of y_o in F with $(\mathbb{R}^n,0)$, $n = \dim F$, we get a homomorphism

$$\text{hol} : \pi_1 L \to G_n^r(0)$$

which is defined up to conjugation.

From now on, by the holonomy of any leaf we always understand a homomorphism in $G_n^r(0)$ (obtained as above).

With our new terminology we can formulate Haefliger's theorem telling us that for a proper leaf L the foliation in a neighbourhood of L is characterized by the holonomy of L.

The proof consists of an application of 1.3.5 and 2.1.3.

2.1.7. - Theorem. (Haefliger [Ha], Siebenmann [Si], Harrison [Har])

Let (M,F) and (M',F') be foliated manifolds of class C^r and codimension n and let L be a proper leaf of both F and F' such that the corresponding holonomy representations are conjugate in $G_n^r(0)$. Then the germs of F and F' near L are conjugate by a diffeomorphism which is the identity on L.

It is natural to ask whether, in the above theorem, one can always find a homeomorphism between saturated representatives of $g(F,L)$ and $g(F',L)$. Exercise 2.1.11, vi) shows, however, that this is far from being true.

The problem of finding arbitrary small saturated neighbourhoods of a leaf takes us back to one of the first papers on foliations, namely Reeb's thesis [Re]. This problem is solved there for compact leaves with finite holonomy group (of differentiable foliations) in the following way.

2.1.8. - Theorem. - (Reeb's local stability theorem)

Let (M,F) be a foliation and $L \in F$ a compact leaf with finite holonomy group. Then there exists a saturated neighbourhood W of L in M and a map

$$p : W \to L$$

which, when restricted to any leaf L' in W, is a finite covering map.

In particular, W consists of compact leaves.

We will get Reeb's theorem as a corollary of the following theorem.

2.1.9. - Theorem.- Let $(M,\mathcal{F})$ be a foliation of class C^r and codimension n and let $L \in \mathcal{F}$ be a compact leaf with finite holonomy group hol(L) . Then there exists a saturated neighbourhood W of L in M and a C^r map

$$p : W \to L$$

such that $(W,p,L,\mathcal{F}|W)$ is a foliated bundle of rank n and structure group hol(L) .

Proof : Let $q : \tilde{L} \to L$ be the covering of L with hol(L) as group of covering translations. Then $\tilde{L}$ is compact. We take an unwrapping $\psi_L = (E,p,L,\mathcal{F}_L)$ of $\mathcal{F}$ near L with C^r immersion $\alpha : E \to M$ which, by the compactness of L , may even be supposed to be an embedding, according to 2.1.2.

Clearly, the germ of the induced pseudobundle $q^*\psi_L$ near $\tilde{L}$ is trivial; compare 1.3.6. Hence, by 1.2.3, $q^*\psi_L$ contains a product of rank n . We conclude that ψ_L contains a foliated bundle of rank $\mathbb{R}^n$ and structure group hol(L). Since α is an embedding, the assertion is proved. □

2.1.10.- Remark.- The previous theorem applies obviously when $\pi_1 L$ is finite.

2.1.11. - Remarks and exercises.- i) The existence of a leaf with non-trivial holonomy implies that the foliation cannot be a fibration.

ii) On the other hand, the flow lines of an irrational flow on the torus constitute a foliation $\mathcal{F}$ without holonomy (i.e. each leaf of

F has trivial holonomy), but F is not a fibration.

iii) A foliation of codimension one without holonomy on a compact manifold is a fibration provided that it has a compact leaf.

iv) Describe the unwrapping construction for the boundary leaf of a Reeb component (for the torus leaf of a Reeb foliation on S^3).

v) Calculate the holonomy of the leaves of a Reeb component (of the Reeb foliation on S^3).

vi) Let Σ be the orientable closed surface of genus two. Our intention is to construct a foliated bundle $\xi = (M,p,\Sigma)$ with fibre the interval such that the transverse foliation on M admits a proper leaf L with trivial holonomy, but with no saturated neighbourhood of L a product. We proceed in several steps.

a) Let $I_1 = [-1,1]$. Show that there exists a sequence $\{f_i\}_{i \in \mathbb{N}}$ of elements of $\mathrm{Homeo}_+(I_1)$ with the following properties:

If $\mathrm{supp}\, f_i$ denotes the <u>support</u> of f_i, i.e.

$$\mathrm{supp}\, f_i = \mathrm{cl}\{ t \in I_1 \mid f_i(t) \neq t \},$$

then

(1) $\mathrm{supp}\, f_i$ is an interval,

(2) $\mathrm{supp}\, f_i \subset \mathrm{supp}\, f_{i+1}$,

(3) $\bigcup_i \mathrm{supp}\, f_i = \overset{\circ}{I}_1$,

(4) $\bigcap_i \mathrm{Fix}(f_i) = \{-1,1\}$.

b) Denote by $G \subset \mathrm{Homeo}_+(I_1)$ the subgroup generated by $\{f_i\}$. Prove that if $G(t)$ is the orbit of G through $t \in I_1$ then $G(-1) = -1$, $G(1) = 1$, and $\overline{G(t)} \supset \{-1,1\}$ for each $t \in \overset{\circ}{I}_1$.

c) If G_t is the group of germs at t of elements of the isotropy group G_t then $G_{-1} = G_1 = \mathrm{id}$.

d) Show that there exists a cyclic covering Σ' of Σ and a representation

$$H' : \pi_1\Sigma' \to G$$

such that, if $(M',\mathcal{F}')$ is the suspension of H', and I_1 is canonically identified with the fibre over the base point $x_o \in \Sigma'$, then the leaf L_1 through 1 is contained in $\bar{L}_t$, for every $t \in \mathring{I}_1$.

What are the saturated neighbourhoods of L_1 ?

e) Let $I_2 = [-2,2]$. The homeomorphisms f_i extend by the identity to elements of $\mathrm{Homeo}_+(I_2)$. Let $g \in \mathrm{Homeo}_+(I_2)$ with $g(1) = -1$ and $g(t) < t$ for $t \in \mathring{I}_2$. Put

$$h_i = g^{-i} \circ f_i \circ g^i ,$$

$$k_i = h_i \circ \ldots \circ h_1 , \quad i \in \mathbb{N} .$$

Show that $\{k_i\}$ converges uniformly to $k \in \mathrm{Homeo}_+(I_2)$.

f) Now let $G \subset \mathrm{Homeo}_+(I_2)$ be the subgroup generated by g and k. Show that there exists a representation

$$H : \pi_1\Sigma \to G$$

such that, if $(M,\mathcal{F})$ is the suspension of H, and I_2 is identified with the fibre over $x_o \in \Sigma$, then the leaf L_1 through 1 is proper and without holonomy.

Furthermore, $\bar{L}_t \supset L_1$ for each $t \in \mathring{I}_2$.

g) Show that the above construction can be made C^r, $r \leqslant \infty$. Moreover, when $r \geqslant 2$ (also for $r < 2$?) all leaves of $(M,\mathcal{F})$ different from L_1, L_2, L_{-2} are dense in M.

2.2. *Holonomy and foliated cocycles ; leaves without holonomy.*

We now give an alternative description of leaf holonomy, in the hope of making this important concept still clearer. This description makes use of the very definition of foliation (by means of foliated cocycles) and so, possibly, is more common than our previous one. Furthermore, it enables us to prove the announced result on the set of leaves without holonomy, (see 2.2.6).

2.2.1. - *Construction and remarks*.- Suppose that the foliation (M,F) of class C^r and codimension n is given by the foliated cocycle $C = (\{(U_i,f_i)\},\{g_{ij}\})$ where $\mathcal{U} = \{U_i\}$ is a nice covering of (M,F) ; see 1.2.4.

i) For each i, we denote by $Q_i \subset \mathbb{R}^n$ the space of plaques of U_i, i.e. $Q_i = f_i(U_i)$. As $\mathcal{U}$ is nice, the maps $g_{ij}(x)$ determine a local diffeomorphism

$$g_{ij} : Q_j \to Q_i \quad ,$$

(with maximal domain), according to 1.2.6. We put $Q = \coprod_i Q_i$. Then $\{g_{ij}\}$ generates a pseudogroup P of local C^r diffeomorphisms of Q, i.e. every element of P is a finite composition of elements g_{ij}.

ii) For $y_o \in Q$, we denote by P_{y_o} the isotropy pseudogroup of P in y_o (i.e. the sub-pseudogroup of P consisting of all elements g with $g(y_o) = y_o$). Its corresponding group of germs G_{y_o} is then identified with a subgroup of $G^r_n(0)$ which is defined up to conjugation, depending on the choice of the maps f_i.

iii) It is not hard to see that equivalent foliated cocycles yield isotropy pseudogroups whose corresponding groups of germs are conjugate.

2.2.2. - *Geometric description of P and P_{y_o}*.

We now want to describe the elements of P_{y_o} as a kind of "Poincaré map" obtained by lifting a closed path in the leaf L through $x_o \in f_i^{-1}(y_o)$, $y_o \in Q_i$, to the nearby leaves.

i) We first realize Q geometrically as a submanifold of M

(possibly with selfintersections) by choosing for each i a transverse section

$$\hat{f}_i : Q_i \to U_i$$

of f_i with image $\hat{Q}_i$.

If $U_i \cap U_j \neq \emptyset$ then, by 2.2.1, the map g_{ij} induces a local diffeomorphism

$$\hat{g}_{ij} : \hat{Q}_j \to \hat{Q}_i \qquad .$$

Thus P yields a pseudogroup $\hat{P}$ of local C^r diffeomorphisms of $\hat{Q}$, a so-called <u>geometrical realization of</u> P (or a <u>geometrical holonomy pseudo-group of</u> F).

ii) Let $U_i \cap U_j \neq \emptyset$ and let U_{ij} be a distinguished open cube containing $cl(U_i \cup U_j)$. Denote by $\hat{D}_j$ the domain of $\hat{g}_{ij}$. There exists a continuous map

$$h_{ij} : \hat{D}_j \times I \to M$$

such that

(1) $h_{ij}(x,0) = x$, $h_{ij}(x,1) = \hat{g}_{ij}(x)$,

(2) $h_{ij}(\{x\} \times I)$ is a path in the leaf L_x , cf. fig. 4.

This means that $\hat{g}_{ij}$ is obtained by lifting a path in L_{x_o} to the nearby leaves. In a similar way we can describe any $\hat{g} \in \hat{P}$.

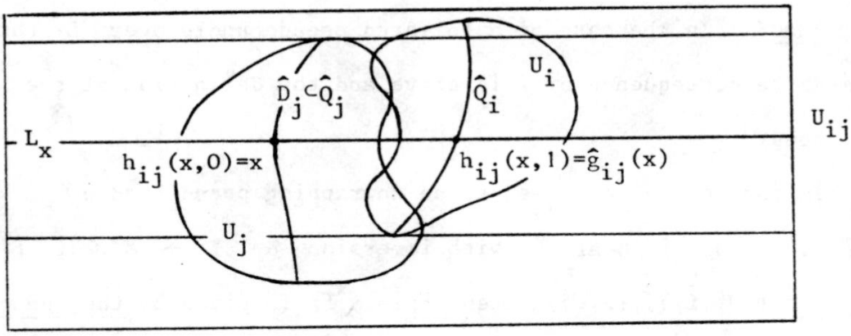

Figure 4

iii) In the particular case of a foliated pseudobundle

$\psi = (E,p,L,\mathcal{F})$ we can take the open cubes U_i to be bidistinguished, with $U_i \cap L$ a single plaque P_i, and the $\hat{Q}_i$'s to be pairwise distinct fibres F_i of p.

Then $\hat{g}_{ij}$ is the local translation of F_j to F_i along some curve c_{ji} in $P_i \cap P_j \subset L$. This shows that any element of $\hat{\mathcal{P}}$ is a local translation.

Conversely, if c is a path in L joining two points of $\hat{Q}$ then c is homotopic to a finite composition

$$c_{i_1 i_2} * c_{i_2 i_3} * \dots * c_{i_{s-1} i_s}$$

where $c_{i_{\sigma-1} i_\sigma}$ lies in $P_{i_{\sigma-1}} \cap P_{i_\sigma}$. Thus the germ of T_c at $F_{i_1} \cap L$ equals the germ of

$$T_{c_{i_1 i_2}} \circ \dots \circ T_{c_{i_{\sigma-1} i_\sigma}} \in \hat{\mathcal{P}} \quad .$$

The relationship between the isotropy pseudogroup $\mathcal{P}_{y_o}$ and the holonomy group of the leaf L of $\mathcal{F}$ passing through the point $x_o = \hat{f}_i^{-1}(y_o)$ is now provided by the following theorem.

2.2.3. - Theorem.- (Notation as in 2.2.1 and 2.2.2)

<u>The group of germs at y_o of $\mathcal{P}_{y_o}$ is canonically isomorphic to the holonomy group</u> hol(L) <u>of the leaf</u> L <u>of</u> $\mathcal{F}$ <u>passing through</u> x_o.

Proof : In the case of a foliated pseudobundle over L the result is an immediate consequence of iii) above and the definition of the holonomy group.

In the general case we use an unwrapping pseudobundle $\psi_L = (E,p,L,\mathcal{F}_L)$ of $\mathcal{F}$ near L with immersion $\alpha : E \to M$. If $\mathcal{F}$ is given by $C = (\{U_i,f_i)\},\{g_{ij}\})$ then $\mathcal{F}_L = \alpha^*\mathcal{F}$ is given by the <u>induced</u> foliated cocycle

$$\tilde{C} = \alpha^* C = (\{(V_{i_j}, f_i \circ \alpha)\} \{g_{ij}\})$$

whose underlying covering is formed by the components of $\{\alpha^{-1}(U_i)\}$.

Let $\hat{P}$ and $\tilde{P}$ be the holonomy pseudogroups of F and F_L with respect to C and $\tilde{C}$, respectively. Then α induces a homomorphism of the isotropy pseudogroups $\tilde{P}_o$ and $\hat{P}_{x_o}$ which is an isomorphism on the corresponding groups of germs. □

2.2.4. - *Remark*.- The preceding theorem justifies calling the isotropy pseudogroup P_{y_o} the holonomy pseudogroup of the leaf L (with respect to the foliated cocycle C).

2.2.5. - *Remark*.- Yet another, but equivalent, definition of leaf holonomy, using a sheaf theoretic approach, can be found in Haefliger's article [Ha].

We conclude this section with a description of the subset of a foliated manifold (M,F) which is formed by the leaves of F with trivial holonomy. As we shall see, this set is always dense in M. More precisely, we have the following even stronger result due to Epstein - Millett - Tischler and the first author; see [EMT] and [Hec]. Here a subset of M is called residual if it is the intersection of countably many dense open subsets. Note that, by the Baire property of the manifold M, every residual subset of M is dense in M.

2.2.6. - *Theorem*.- Let (M,F) be a foliated manifold. The subset of M which is formed by the leaves with trivial holonomy is residual in M.

Moreover, there are uncountably many leaves with trivial holonomy.

Proof : Let (M,F) be given by the foliated cocycle $C = (\{(U_i,f_i)\},\{g_{ij}\})$ where $\{U_i\}$ is a nice covering. Let $\hat{Q} = \coprod_i \hat{Q}_i$ and let $\hat{P}$ be the geometrical holonomy pseudogroup (see 2.2.2). Note that $\hat{P}$ is countable, because $\{U_i\}$ is countable. For $\hat{g} \in \hat{P}$ we denote by

$Fix(\hat{g}) \subset \hat{Q}$ the closed subset of fixed points of $\hat{g}$. We put

$$\partial Fix(\hat{g}) = Fix(\hat{g}) - int(Fix(\hat{g}))$$

and

$$B = \bigcup_{\hat{g} \varepsilon \hat{P}} \partial Fix(\hat{g}) .$$

By the definition of B , we see immediately that $x \varepsilon \hat{Q}_i - B$ if and only if the leaf of F through x has trivial holonomy. But B is a countable union of closed nowhere dense subsets of $\hat{Q}$. Thus $\hat{Q} - B$ is residual in $\hat{Q}$.

Now if H denotes the subset of M which is formed by the leaves of F with non-trivial holonomy then

$$H \cap \hat{Q} = B .$$

We conclude that the complement of H in M is the intersection of countably many dense open sets. □

2.2.7. - Remarks.- (See Epstein - Millett - Tischler [EMT])

i) In the above theorem the hypothesis that M has a countable basis of its topology is essential. Indeed, one can construct a foliation F on some non-paracompact 3 - manifold such that

(1) F consists of a single leaf L ,

(2) the holonomy of L is non-trivial.

ii) There are well-known examples of foliations showing that in general the set of leaves with trivial holonomy cannot be expected to be open. See the next exercises.

2.2.8. - Exercises.- i) Let the diffeomorphism $A : T^2 \to T^2$ be given by the matrix $A \varepsilon SL(2;\mathbb{Z})$ with trace greater than two. Then A has two irrational real eigenvalues α and β . The lines parallel to the eigenspace of α induce a linear foliation on T^2 . The corresponding product foliation on $T^2 \times \mathbb{R}$ is preserved by A and so yields a

2-dimensional foliation $\mathcal{F}$ on the mapping torus $M_A = (T^2 \times \mathbb{R}) / A$. Recall that M_A is nothing else than the total space of the suspension of the representation $\pi_1 S^1 \rightarrow \mathrm{Diff}^\infty(T^2)$ which takes $1 \in \pi_1 S^1 = \mathbb{Z}$ to A. Show that

a) the leaves of $(M_A, \mathcal{F})$ are either planes or cylinders, i.e. homeomorphic to $\mathbb{R}^2$ or $S^1 \times \mathbb{R}$, respectively,

b) all leaves are dense in M_A,

c) there are countably many cylinders all of which have non-trivial holonomy.

d) Conclude that the set of leaves of $\mathcal{F}$ without holonomy has empty interior.

ii) Let Σ be the closed orientable surface of genus two and let G be the subgroup of $\mathrm{Diff}^r_+(S^1)$ which is generated by two diffeomorphisms f and g.

a) Show that there exists $H : \pi_1 \Sigma \rightarrow \mathrm{Diff}^r_+(S^1)$ with image G. (See also exercise 2.1.11,vi)f)).

b) Show that, for a suitable choice of f and g, the suspension of H has uncountably many leaves with non-trivial holonomy.

3. *Linear holonomy ; Thurston's stability theorem.*

3.1. *Linear and infinitesimal holonomy.*

In I; 3.5 we have introduced the infinitesimal holonomy for foliated surfaces. We now extend this concept to arbitrary foliations.

3.1.1. - *Definition*.- Let $J^s_n(0)$ be the set of s-jets at $0 \in \mathbb{R}^n$, i.e. the elements of $J^s_n(0)$ are equivalence classes of elements of $G^s_n(0)$, where the equivalence relation is given by

$$[f]_0 \in G^s_n(0) \text{ equals } [g]_0 \in G^s_n(0)$$

if f and g have the same derivatives up to order s.

For $s \leq r$ we have a natural projection

$$J^s : G_n^r(0) \to J_n^s(0) .$$

Now let (M,F) be a foliation of class C^r, $r \geq 1$, and codimension n. For $s \leq r$ and $L \in F$, the homomorphism

$$J^s \circ hol : \pi_1 L \to J_n^s(0)$$

is called the <u>infinitesimal holonomy of order</u> s of L.

In particular, the infinitesimal holonomy of order one is referred to as the <u>linear holonomy</u> of L. It is also written as

$$Dhol : \pi_1 L \to GL(n;\mathbb{R}) .$$

Notice that Dhol is an element of $H^1(L;GL(n;\mathbb{R}))$.

3.1.2. - <u>*The normal bundle of a leaf.*</u>- The <u>normal bundle</u> $\nu_L = (N,p,L)$ <u>of a leaf</u> $L \in (M,F)$, i.e. the vector bundle over L induced by the inclusion $\iota : L \to M$ from the normal bundle NF of F, is a foliated bundle. We want to see that its holonomy representation

$$H_{\nu_L} : \pi_1 L \to GL(n;\mathbb{R})$$

is just Dhol.

Firstly we show that ν_L is indeed foliated. Namely, if F is given by the foliated cocycle $C = (\{(U_i,f_i)\},\{g_{ij}\})$ then NF is defined by the cocycle $C' = (\{U_i\},\{g'_{ij}\})$ where

$$g'_{ij} : U_i \cap U_j \to GL(n;\mathbb{R})$$

assigns to $x \in U_i \cap U_j \neq \emptyset$ the derivative of $g_{ij}(x)$ at the point $f_j(x)$.

Now we can easily see that the induced cocycle ι^*C' on L is equivalent to one which is locally constant. Thus ν_L is a foliated bundle. It is a consequence of the very definitions that H_{ν_L} equals Dhol. We have proved:

3.1.3. - <u>*Lemma.*</u>- <u>The holonomy representation</u> H_{ν_L} <u>of the normal bundle of</u> $L \in F$ <u>coincides with the linear holonomy of</u> L.

3.1.4. - *Remark*.- If $\psi_L = (E,p,L,F_L)$ is an unwrapping pseudo-bundle of F near L, where E is a neighbourhood of the zero-section of ν_L, then in general the foliation F_L is not comparable with the foliation induced on E by the transverse foliation of ν_L.

3.1.5. - *Exercises*.- i) a) Calculate the linear holonomy of the Reeb component defined in II; 1.4.4, 1.4.5.

b) Construct a Reeb component all of whose leaves have trivial linear holonomy.

ii) A foliation (M,F) of class C^1 and codimension one is defined by a Pfaffian form ω on M such that $d\omega = \omega \wedge \alpha$ for some 1-form α. Compare II; theorem 2.4.4 and the remark following it.

a) Show that the restriction $\alpha|L$ of α to any leaf L of F is closed.

b) Show that $\mathrm{Dhol}(\gamma) = \exp \int_\gamma \alpha|L$ for any $\gamma \in \pi_1 L$.

3.2. *Thurston's stability theorem*.

The proof of the Reeb stability theorem (see 2.1.8 and 2.1.9) indicates that the conclusion of the theorem holds under somewhat weaker assumptions on the leaf L. We shall give here a generalized version of Reeb's theorem involving the first real cohomology group of L. Roughly speaking, under certain conditions on the holonomy of L the existence of a non-trivial representation of $\pi_1 L$ in $\mathbb{R}$ can be deduced thus showing $H^1(L;\mathbb{R}) \neq 0$. More precisely, we prove :

3.2.1. - *Theorem*.- (Thurston [Th]) Let F be a foliation of class C^1 and codimension n. For each compact leaf L of F at least one of the following possibilities holds :

(1) The linear holonomy of L is non-trivial.

(2) $H^1(L;\mathbb{R}) \neq 0$.

(3) The holonomy of L is trivial.

3.2.2. - *Corollary*.- If $H^1(L;\mathbb{R}) = 0$ and $H^1(L;GL(n;\mathbb{R})) = 0$ then $hol(L) = 0$ and L admits a neighbourhood on which the foliation F induces a product $L \times \mathbb{R}^n$.

It should be pointed out that the theorem does not hold in the C^0 setting (see 3.2.7). Observe also that F need not be transversely orientable.

3.2.3. - *Preliminaries for the proof of 3.2.1*.- We choose an adapted representative $\psi = (E,p,L,F_L)$ of the unwrapping microbundle of L , with holonomy representation

$$H_\psi : \pi_1 L \to H_n^1(0) .$$

Let $\Gamma = \{\gamma_1,\dots,\gamma_s\}$ be a symmetric (i.e. $\gamma \in \Gamma$ implies $\gamma^{-1} \in \Gamma$) set of generators of $\pi_1 L$. Set

$$g_i = H_\psi(\gamma_i) \text{ and } A = \{g_1,\dots,g_s\} ,$$

and denote by P the sub-pseudogroup of $H_n^1(0)$ generated by A (which, in general, is different from the holonomy pseudogroup as defined in 2.2.4). Then each non-trivial $h \in P$ can be written as

(*) $$h = g_{i_\ell} \circ \dots \circ g_{i_1} , \text{ with } g_{i_j} \in A .$$

As in the case of groups, the least ℓ such that h can be written as a product (*) with ℓ factors is called the length of h (with respect to A), denoted $\ell(h)$. By convention, $\ell(id) = 0$.

Now let h and k be two elements of P and let x be such that x and $k(x)$ lie in a cube around 0 which is contained in the domain of h . Applying the mean value theorem to $h - id$, we get

(1) $\| (h \circ k)(x) - k(x) - (h(x) - x) \| \leqslant \| k(x) - x \| \| D(h - id)(z) \|$,

with $z = z(x) = k(x) + t(k(x) - x)$ and suitable $t \in [0,1]$.

Assuming that $Dhol(L) = 0$, we get, for any $h \in P$,

(2) $\lim_{x \to o} \| D(h - id)(x) \| = 0$.

Denote by U the intersection of the domains of the elements of A . Then there is a sequence $S = \{x_n\}_{n \in \mathbb{N}}$ in U , converging to 0 and an element of A , say g_1 , such that

$$m(x_n) = \| g_1(x_n) - x_n \| \geqslant \| g(x_n) - x_n \|$$

for every $g \in A$.

If $hol(L) \neq 0$ then we can choose S so that $m(x_n) \neq 0$ for any n . We then define for each $h \in P$

$$\mathbb{N}_h = \{ n \in \mathbb{N} \mid x_n \in \text{domain } h \}$$

and

$$S_h = \{ \frac{1}{m(x_n)} (h(x_n) - x_n) \}_{n \in \mathbb{N}_h}$$

3.2.4. - *Lemma*.- If $hol(L) \neq 0$ and $Dhol(L) = 0$ then we can choose S such that S_h is convergent for any $h \in P$.

Proof : We first prove that for any sequence S as above the sequence S_h is bounded for any $h \in P$. This is shown by induction on the length $\ell(h)$.

The only element of length 0 is the identity, and S_{id} is constant. For $h \in P$ and $g_i \in A$ we get, by (1),

$$\frac{1}{m(x_n)} \| h \circ g_i(x_n) - g_i(x_n) - (h(x_n) - x_n) \|$$

$$\leqslant \frac{1}{m(x_n)} \| g_i(x_n) - x_n \| \| D(h - id)(z_n) \| .$$

Hence, by (2)

$$\lim_{n\to\infty} \frac{1}{m(x_n)} \| h \circ g_i(x_n) - g_i(x_n) - (h(x_n) - x_n) \| = 0 .$$

This means that the sequence $\{\frac{1}{m(x_n)}(h \circ g_i(x_n) - g_i(x_n))\}$ is bounded. Writing

$$\| h \circ g_i(x_n) - x_n \| \leqslant \| h \circ g_i(x_n) - g_i(x_n) \| + \| g_i(x_n) - x_n \|$$

and applying the induction hypothesis for g_i, we see that $S_{h \circ g_i}$ is bounded.

The pseudogroup P being countable, we may enumerate its elements $h_1, h_2, \ldots, h_i, \ldots$ By the diagonal process, we choose a subsequence of S for which S_{h_i} converges for any i. □

3.2.5. - *Proof of theorem 3.2.1* : Assume that $\mathrm{hol}(L) \neq 0$ and $\mathrm{Dhol}(L) = 0$. With the notations above we set

$$\hat{H}(h) = \lim S_h \quad , \; h \in P .$$

Clearly, $\hat{H}(h)$ depends only on the germ of h at 0, i.e. we have a commutative triangle

$$\begin{array}{ccc} P & & \\ {\scriptstyle\pi}\downarrow & \searrow^{\hat{H}} & \\ \mathrm{hol}(L) & \xrightarrow{H} & \mathbb{R}^n , \end{array}$$

with π the canonical map. Now it remains to show that H is a non-trivial group homomorphism.

We first observe that for the general element of $S_{h \circ k} - S_h - S_k$ we have the following estimate (see 3.2.3,(1)) :

$$\frac{1}{m(x_n)} \| h \circ k(x_n) - k(x_n) - h(x_n) + x_n \|$$

$$\leqslant \frac{1}{m(x_n)} \| k(x_n) - x_n \| \, \| D(h - \mathrm{id})(z_n) \| .$$

By 3.2.3,(2), the right-hand side tends to zero. This shows that H is a homomorphism.

Finally, notice that each element of S_{g_1} has norm one, hence $H(g_1) \neq 0$. □

3.2.6. - *Remark*.- Other proofs of the existence of a non-trivial homomorphism $H : \pi_1 L \to \mathbb{R}$ have been given by Reeb - Schweitzer and Schachermayer in [RS] and by Jouanolou in [Jo].

3.2.7. - *A counterexample of 3.2.1 in C^0*.- Let $G = \widetilde{(PSL(2;\mathbb{R}))}$ be the universal covering of $PSL(2;\mathbb{R}) = SL(2;\mathbb{R}) / \{\pm \mathrm{id}\}$. Since $PSL(2;\mathbb{R})$ preserves the lines through the origin, it acts on $P^1(\mathbb{R})$. Thus G acts on $\mathbb{R} = \widetilde{P}^1(\mathbb{R})$, the universal covering of $P^1(\mathbb{R})$, and hence on $S^1 = \mathbb{R} \cup \{\infty\}$ with a fixed point. This action is, however, only topological.

There exist many discrete subgroups of G with compact quotient. For example, there is $\Gamma \subset G$ such that $M = G/\Gamma$ is a homology sphere. Suspending the representation

$$H : \pi_1 M = \Gamma \to \mathrm{Homeo}(S^1)$$

yields a foliated bundle with one compact leaf L homeomorphic to M but without any saturated neighbourhood of L by compact leaves.

Literature

[Bi] Birkhoff, G.: Lattice theory. Publ. Amer. Math. Soc. XXV (1948)

[Bl] Blumenthal, R.A.: Transversely homogeneous foliations. Ann. Inst. Fourier 29-4, 143-158 (1979)

[CN] Camacho, C. - Neto, A.L.: Teoria geometrica das folheações. I.M.P.A. Rio de Janeiro (1979)

[De] Denjoy, A.: Sur les courbes définies par les équations différentielles à la surface du tore. J. de Math. 9 (11), 333-375 (1932)

[Eh] Ehresmann, C.: Les connexions infinitésimales dans un espace fibré différentiable. Colloque de Topologie, CBRM, Bruxelles, 29-55 (1950)

[EMT] Epstein, D.B.A. - Millet, K.C. - Tischler, D.: Leaves without holonomy. J. London Math. Soc. 16, 548-552 (1977)

[Fe] Fedida, E.: Feuilletages du plan - feuilletages de Lie. Université Louis Pasteur, Strasbourg (1973)

[Ha] Haefliger, A.: Variétés feuilletées. Ann. Scuola Norm. Sup. Pisa (3) 16, 367-397 (1962)

[Har] Harrison, J.: Structure of a foliated neighbourhood. Math. Proc. Camb. Phil. Soc. 79, 101-110 (1976)

[Hec] Hector, G.: Feuilletages en cylindres. In "Geometry and topology" Rio de Janeiro 1976, Springer LN 597, 252-270 (1977)

[He] Herman, M.R.: Sur la conjugaison différentiable des difféomorphismes du cercle à des rotations. Publ. Math. I.H.E.S. 49, 5-234 (1979)

[Her] Hermann, R.: On the differential geometry of foliations. Ann. Math. (3) 72, 445-457 (1960)

[Hi] Hirsch, M.W.: Differential topology. Graduate Texts in Mathematics 33, Springer Verlag (1976)

[HS] Hirsch, M.W. - Smale, S.: Differential equations, dynamical systems, and linear algebra. Academic Press, New York (1974)

[Hir] Hirzebruch, F.: Topological methods in algebraic geometry. Grundlehren 131, Springer Verlag (1976)

[Hu] Husemoller, D.: Fibre bundles. Graduate Texts in Mathematics 20, Springer Verlag (1975)

[Jo] Jouanolou, J.: Une preuve élémentaire d'un théorème de Thurston. Topology, 17, 109 - 110 (1978)

[Kn] Kneser, H.: Reguläre Kurvenscharen auf den Ringflächen. Math. Ann. 19, 135 - 154 (1924)

[La] Lang, S.: Analysis II. Addison - Wesley Publ. Comp. (1969)

[Mi] Milnor, J.: Topology from the differentiable viewpoint. The University Press of Virginia (1965)

[MS] Milnor, J. - Stasheff, J.D.: Characteristic classes. Ann. Math. Studies 76, Princeton University Press (1974)

[Ni] Nitecki, Z.: Differentiable dynamics. The M.I.T. Press, Cambridge, Mass. (1971)

[Po] Poincaré, H.: Oeuvres complètes, tome I, Gauthier - Villars, Paris (1928)

[Pu] Pugh, C.: The closing lemma. Amer. J. Math. 89, 956 - 1009 (1967)

[Re] Reeb, G.: Sur certaines propriétés topologiques des variétés feuilletées. Actualités Sci. Indust., Hermann, Paris (1952)

[RS] Reeb, G. - Schweitzer , P.A.: Un théorème de Thurston établi au moyen de l'analyse non standard. In "Differential topology, foliations and Gelfand - Fuks cohomology". Springer LN 652, p. 138 (1978). Addendum by W. Schachermayer: Une modification standard de la démonstration non standard de Reeb et Schweitzer. ibidem 139 - 140

[Rei] Reinhart, B.L.: Foliated manifolds with bundle - like metrics. Ann. Math. (2) 69, 119 - 132 (1959)

[Ro] Rosenberg, H.: Un contre - exemple à la conjecture de Seifert. Séminaire Bourbaki, Exposé 434, Springer LN 383, 294 - 306 (1973)

[Sa] Sacksteder, R.: Foliations and pseudo - groups. Amer. J. Math. 87, 79 - 102 (1965)

[Sc] Schwartz, A.: A generalization of a Poincaré - Bendixson theorem to closed two - dimensional manifolds. Amer. J. Math. 85, 453 - 458 (1963)

[Si] Siebenmann, L.C.: Deformation of homeomorphisms on stratified sets. Comm. Math. Helv. 47, 123 - 163 (1972)

[Sie] Siegel, C.L.: Notes on differential equations on the torus. Ann. Math. 46, 423 - 428 (1945)

[St] Steenrod, N.: The topology of fibres bundles. Princeton Math. Series 14, (1951)

[Ste] Sternberg, S.: Lectures on differential geometry. Prentice - Hall, Englewood Cliffs (1964)

[Th] Thurston, W.P.: A generalization of the Reeb stability theorem. Topology 13, 347 - 352 (1974)

[Wo] Wolf, J.A.: Spaces of constant curvature. Publish or Perish Inc., Berkeley (1977)

This is the list of cited literature. For further references and information consult B. Lawson's survey articles :

1. Foliations. Bull. Amer. Math. Soc. 80, 369 - 418 (1974)
2. The quantitative theory of foliations. CBMS, Reg. Conf. Ser. Math. 27, Amer. Math. Soc., Providence, Rhode Island (1977)

Glossary of notations

$\mathbb{R}^n$	n-dimensional euklidean space 2
T^n	n-dimensional torus 3
S^n	n-dimensional sphere 3
M	manifold 5
$I(X,x)$	Index of the vector field X at the point x 5
K^2	Klein bottle 9
(Σ,F)	foliation on the surface Σ 12
F	foliation 12,148
L_x	leaf passing through the point x 13,153
$\partial\Sigma$	boundary of the surface Σ 14
$F^{\pitchfork}$	foliation transverse to F 18
$\chi(\Sigma)$	Euler characteristic of Σ 25
$Diff^r(M)$	group of C^r diffeomorphisms of M 29
Homeo(M)	group of homeomorphisms of M 29
$Diff^r_+(M)$	group of orientation preserving C^r diffeomorphisms of M 29
I	interval 29
$D^r(S^1)$	group of $\mathbb{Z}$-periodic elements of $Diff^r_+(\mathbb{R})$ 30
$\bar{R}_\alpha$	rotation of S^1 through α 30
$F\|U$	restriction of the foliation F to the subset U 34,154
$g(F,L)$	germ of the foliation F near the leaf L 34
$G^r(\mathbb{R},0)$	group of germs at 0 of C^r diffeomorphisms which are defined in a neighbourhood of 0 and keep 0 fixed 37
$G^r(\mathbb{R}^+,0)$	Group of germs at 0 of C^r diffeomorphisms which are defined in a neighbourhood of $0 \in \mathbb{R}^+$ and keep 0 fixed 37
$G^r_+(\mathbb{R},0)$	subgroup of $G^r(\mathbb{R},0)$ consisting of those germs which are represented by orientation preserving diffeomorphisms 37
hol	holonomy representation 38,207
hol(L)	holonomy group of the leaf L 38,207
$\bar{A}$, cl(A)	closure of A 45
$\overset{\circ}{A}$, int(A)	interior of A 45
$\mathcal{M}$	minimal set 46
$\rho(f)$	rotation number of $f \in D^0(S^1)$ 71
Fix(f)	fixed point set of the homeomorphism f 89

Per(f)	set of periodic points of f 89
$F^1(\Sigma)$	foliations on Σ with C^1 tangent vector field 95
$F^1(\xi)$	subspace of $F^1(\Sigma)$ consisting of all foliations with projectable tangent vector field 96
$F^1_\partial(\xi)$	subspace of $F^1(\xi)$ consisting of all foliations which are tangent to the boundary 97
$C(F)$	set of circle leaves of the foliation F 96
Dhol	linear holonomy 104,218
ξ	fibre bundle 113
C	cocycle, foliated cocycle 116,150
$f^*\xi$	induced fibre bundle 121
$\mathrm{Homeo}(F)^\delta$	Homeo(F) endowed with the discrete topology 124
M^δ	manifold M endowed with the leaf topology 127,153
(M,F)	foliation on the manifold M 148
H^r_n	pseudogroup of local diffeomorphisms of $\mathbb{R}^n$ 150
H^r_{n+}	pseudogroup of orientation preserving elements of H^r_n 150
f^*F	induced foliation 159
TM	tangent bundle of M 163
TF	tangent bundle of F 164
NF	normal bundle of F 164
[X,Y]	Poisson bracket of the vector fields X and Y 165
ψ	foliated pseudobundle 186
μ	foliated microbundle 189
$\bar{f}^*\mu$	induced microbundle 190
ψ_L	unwrapping of the leaf L 205
μ_L	unwrapping microbundle of the leaf L 205
$G^r_n(0)$	group of germs at 0 of local C^r diffeomorphisms of $\mathbb{R}^n$ which are defined in a neighbourhood of 0 and keep 0 fixed 194
$H^r_n(0)$	pseudogroup consisting of all representatives of elements of $G^r_n(0)$ 194

Index